Mohammad Mursaleen Butt
Sheikh Nazir Ahmad

Modelling Smart Systems

LAP LAMBERT Academic Publishing

Cover image: www.ingimage.com

Publisher:
LAP LAMBERT Academic Publishing
is a trademark of
International Book Market Service Ltd., member of OmniScriptum Publishing Group
17 Meldrum Street, Beau Bassin 71504, Mauritius

Printed at: see last page
ISBN: 978-613-9-45441-9

Zugl. / Approved by: Mohammad Mursaleen, NIT Srinagar 2018

MODELLING SMART SYSTEMS

By

Dr. MOHAMMAD MURSALEEN BUTT

&

Prof. SHEIKH NAZIR AHMAD

Professor

Mechanical Engineering Department

National Institute of Technology Srinagar

Kashmir India

CONTENTS

LIST OF FIGURES

NOMENCLATURE

$\vec{a}$	Displacement vector from negative charge to positive charge
{a}	Global degrees of freedom matrix
$\vec{B}$	Gravity force:
b_{mn}	Electric susceptibility
C_{mnkl}	Elastic constants
C_θ	Thermal Constant
C	Specific heat
$\acute{C}_{LN}$	Scalar product of gradient vectors
$\vec{D}$	Electric displacement
dA	Elemental surface area
dV	Elemental volume
$\overrightarrow{dS}$	Displacement vector along arbitrary path
$\vec{E}$	Electric field
$\vec{e_i}$	Identity vector
e_{mnk}	piezoelectric constant
$\vec{F}$	Force vector
$F_j^l(z)$	Lagrangian interpolation function
$\vec{f}$	Force per unit volume of a polarized medium
$\hat{G}_{LN}$	Green strain tensor
I_j	Element of beam
K_{ij}	Thermal conductivity tensor
[K]	Global stiffness matrix
$[K_1]$ *and* $[K_2]$	Augmented stiffness matrices due to distributed nonlinear force and moment
$\vec{n}$	Identity vector
$\vec{p}$	Dipole moment
$\vec{P}$	Polarization vector
q*	Test charge
q_i	Point charge
q^b	Surface charge density
Q_i	Heat flux
$\vec{R}$	Position vector
$\vec{R}_L$ & $\vec{R}_E$	Undeformed & deformed position vector of point P_0

NOMENCLATURE

T_i	Boundary traction applied
U	internal energy per unit mass
u, v, w	Displacements in undeformed coordinates in x, y and z direction respectively
$\vec{v}$	Velocity vector
$\hat{W}_E$	Rate of work done per unit volume
$\vec{\omega}$	dual vector
W_L	Scalar product of gradient vector and electric field vector
x, y, z	coordinate system
X_i	Undeformed coordinates
y_l	Deformed coordinates
$y_{i,L}$ & $X_{L,j}$	deformation gradients
( ),i	Derivative w. r. t i
α_{mn}	Thermoelastic constant
γ	Rate of heat supplied from internal source per unit mass
∇	Gradient symbol
ε_0	Permittivity of free space
ε	Permittivity
χ_e	Electric susceptibility
η_m	Pyroelectric constant
λ	Entropy
θ	Temperature difference
ϵ_{ij}	Strain tensor
σ_{lm}	Stress tensor
ρ_m	density
ϕ	electric potential
ρ^f	Free charge
ρ^b	Bound charge density
χ	Helmholtz free energy
δW	Infinitesimal work done
$\delta()$	Variation
$\vec{\tau}_p$	moment density
$\vec{\pi}$	Polarization per unit mass

Chapter-I

SMART MATERIALS AND SYSTEMS

1.1 Introduction

Advances in materials have led to the demand for portable and high-performance control design of devices and have given birth to use of intelligent/smart materials and/or structures. The evolution of substantial and economical high-achievement materials and systems are essential for the economic health of a nation, as human infrastructure cost amounts to a considerable chunk of national capital. In order to take care of the issue of degenerating structure, research of smart materials becomes mandatory. Use of such materials will not only improve the performance of the structure, reduce the maintenance costs, ensure or confirm that the structure is sustainable in future but will as well put forth the utilization of smart materials for most favourable achievement or performance and secured infrastructure designs particularly in seismic and other naturally hazardous prone zones.

Smart is the new tech-savvy because smart materials may work completely on their own as a part of the larger smart system. In principle, these systems interconvert the fundamental energy forms viz electric, sound, thermal, mechanical, magnetic etc. Smartness may range from passive to sensible depending upon its responsiveness. Smartness makes certain that under a variety of environmental conditions the system is not only giving the most favourable performance but is smart enough to get suitably actuated to tackle abnormal loads as and when required much like its biological analogue that is a human body.

A traditional system is designed for a specific purpose and performance while these smart systems accommodate to changing environments, offer self-repair or even alert us. The switch to these materials is evident for their real-time response processing, portability, functionality, minimal power requirement, better reliability, scope for innovation, integration and of course miniaturisation, such systems have inbuilt sensor, activator and a control mechanism efficient enough to sense an impetus, reacting to the stimuli in a predestined way and art of reverting back to its original state. In addition to all these smart materials ensure

Damage Arrest: - These can instantaneously produce compressive stresses around the cracks and arrest their propagation.

Shock Absorbs: - These can distinguish between the static and shock loads and can accordingly generate resistance against shock.

Auto-Heal:-These materials are capable of auto-repair.

The beauty of these materials is such that the very same material can act as a sensor and actuator as well. Whereas the sensor by piezoelectric or electrostrictive effect does involve the transformation of mechanical variable force or displacement into quantifiable electrical quantity, alternatively the actuator via converse or indirect piezoelectric effect involves a conversion of the electrical signal into force or useful displacement. The qualities of performing both actuation and sensing are uniquely being exhibited by ferroelectric materials.

1.2 Ferroelectricity and Ferroelectric Materials

Ferroelectricity: An impetuous electric polarization exhibited by certain materials and that polarization is revertible under execution of an electric field, traditionally the characteristic of dielectric crystals. When electric field is applied, the charged particles of crystal tend to dislocate being loaded by the electrostatic forces. The direction of polarization can be reverted with a change in dimensions when being subjected to the coercive level of an electric field opposite to the polarized orientation, during this instability phase abrupt shape changes occur with changes from one crystal state to another and electrostatic forces rearranging the location of negative and positive charges. The cluster polycrystalline is composed of several fundamental crystals in a way that polarized direction of each and every crystal differ from the other that constitute the aggregate, resulting in a polycrystalline which is not polarized at the global level. On subjecting the material to an electric field at the coercive level each individual crystal switches its polarized direction along the applied field and material gets polarized, on withdrawing the electric field the material retains a degree of polarization and remains polarized with individual crystal poling orientations mostly aligned along the coercive field direction. For electric field variation in a cyclic fashion, the polarisation manifest hysteresis loop implying a loss of energy.

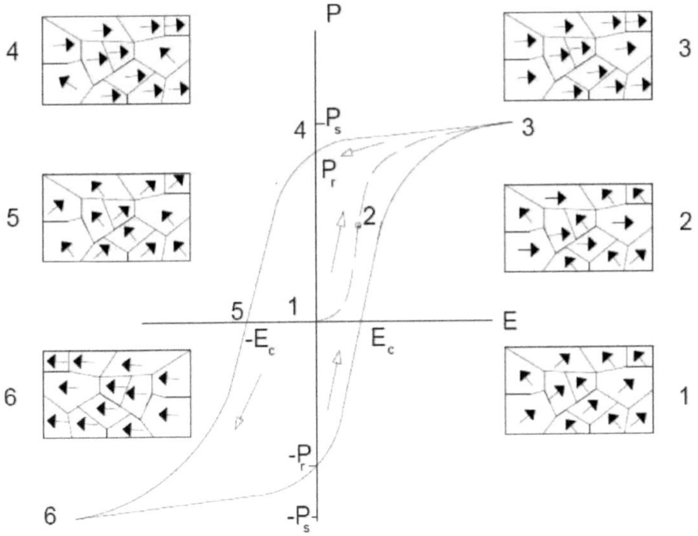

Figure 1.1: Hysteresis in Ferroelectric Ceramics

The hysteresis cycle is illustrated by analysing the outcome of electric field on polarisation. For a crystal with undirected and unarranged haphazardly positioned dipoles being in unpoled condition, implementing an electric field starts to align the dipoles with the field as illustrated for coordinates1 and 2 of the Fig, 1.1, and maximum polarisation Ps is obtained at 3 on increasing the electric field. On withdrawing the electric field, a residual polarisation P_r remains ordinate 4, and the material gets polarised permanently due to interactions between dipoles. Reversing the electric field reverses the direction of dipoles. The negative coercive field E_c randomizes the dipole directions coordinates 5 and depolarizes the material, enhancing the electric field further in the negative direction ensures saturation in opposite direction and polarisation at Point 6 is a mirror image of polarization of point 3. Withdrawing the electric field from point 5 gives polarisation that is exactly opposite to polarization obtained when a field was withdrawn from point 3. Alternating the field leads to the formation of a hysteresis loop with the area of the loop and work requirement for swapping the polarisation by 180 degrees being interrelated. The ferroelectric behaviour of revertible and spontaneous polarization and formation of loop for cyclic variations of polarisation and electric field rely on temperature. The relationship of polarisation and an electric field being non-linear in the ferroelectric region with the material exhibiting hysteresis. Nonetheless in excess of a definite temperature known as Curie temperature, the phenomenon of formation of hysteresis vanish and polarisation vector is linearly related to the electric field vector and the dielectric is said to be in a paraelectric state.

Ferroelectricity is thus an electrical property wherein a material gets polarized upon experiencing a nearby electric field. Polarization is displacement of certain ions in a structure in response to an external field. In dielectrics (insulators) all charges are attached to a specific atom or molecule in a tight leash all they can do is move a bit within the atoms/molecules. Ferroelectric materials retain this state even after the removal of an electric field, they own the average dipole moment per unit cell (spontaneous and abrupt polarization).The direction of polarisation is reversible as a result of the reason that ferroelectric polar structure is, in fact, a slightly distorted nonpolar structure. This distortion leads to nonlinear dielectric behaviour. These crystals though do not have an inversion centre, yet they have a specific polar axis. They belong to a group of polar dielectrics. Their spontaneous polarization can be toggled upon application of an electric field and is exhibited in an explicit manner in P-E hysteresis loop. These undergo a phase transition from paraelectric phase at high temperature into ferroelectric phase at low temperature through Curie temperature. Above the Curie temperature ferroelectric crystals lose their ferroelectric properties and tend to become paraelectric materials with centrosymmetric crystallography and with barely any spontaneous polarization. Ferroelectrics have very high dielectric constants at relatively low applied electric field frequencies. In order to improve properties of ferroelectric ceramics we chemically modify them by doping which can be either addition of donor dopants to create cation vacancies with impurities like in a PZT, the replacement of Pb^{2+} with higher valence dopants like Bi^{3+} or La^{3+} and $Ti^{4+}(Zr^{4+})$ with Nb^{5+} or Ta^{5+}, which lead to reducing aging, lower coercive fields, increased dielectric constant, high piezoelectric coupling etc., there crystals after being doped with higher valence dopants are known as soft ferroelectrics. On the contrary when being doped with lower valence dopants or addition of acceptor dopants to create anion vacancies known as hard ferroelectrics like Pb^{2+} with Na^{+} or K^{+} and Ti^{4+} (Zr^{4+}) with Mg^{2+} or Fe^{3+} for PZT, which would exhibit lower electrical resistivity, lower dielectric constants, lower dielectric losses and higher coercive fields. The soft ferroelectrics result with having low melting or decomposition temperatures, water soluble and mechanically soft. The hard ferroelectrics developed at high temperatures are water-insoluble and mechanically hard. Another classification of ferroelectrics is uniaxial ferroelectrics and multiaxial ferroelectrics, uniaxial ferroelectrics crystals polarise along one axis only and multi-axial ferroelectrics polarise along various axes. In general, hard ferroelectrics are multiaxial and soft are uniaxial.

Soft ferroelectrics normally have a complex molecular structure and include Sodium potassium tartrate tetrahydrate ($NaKC_4H_4O_64H_2O$) or Rochelle salt, Potassium dihydrogen phosphate (KH_2PO_4 or H_2KO_4P) and Guanidine aluminium sulphate hexahydrate ((CN_3H_6) $Al(SO_4)_2 \cdot 6H_2O$) or GASH. The structural characteristics of hard (oxide) ferroelectrics are well-organized and crystallize usually into perovskite structure, labelled after the mineral (Calcium Titanate). Simple cubic structure is the high-temperature form of most of the oxides of the type ABO_3.

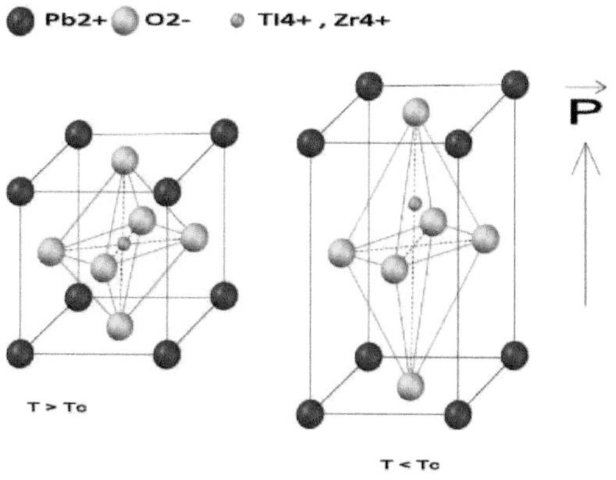

Figure 1.2: Perovskite Structure

1.3 Pyroelectricity

In some crystals, temporary voltage appears upon being cooled or heated. The change in temperature changes the positions of atoms to some degree within the crystal leading to polarization which results in generating a potential across the crystal. It is found in crystals like Gallium nitride (GaN), Caesium nitrite ($CSNo_3$), polyvinyflourides (PVF), tourmaline etc. First reference of pyroelectricity comes from writings of Theophrastus in

314 BC. However properties of tourmaline were re-described by Johann George Schmidt in 1707. David Brewster bestowed the effect in 1824, the name it is known by today. Woldemar Voigt in 1897 and subsequently William Thomson in 1878 developed theories for explaining the phenomenon of pyroelectricity. Every ferroelectric material is pyroelectric but the converse is not true.

1.4 Electrostriction and Piezoelectricity

Electrostriction: - A second order effect that occurs in dielectric materials and produces a change in shape (deformation) of the material on application of an electrical field. The relation between Electrostrictive strain and vector field being quadratic. This process is irreversible, occurs in all dielectric materials, spawned by electrical domains that are randomly oriented, the electrical domains get aligned by applied electric field ensuring elongation in the field direction and reducing the thickness of materials orthogonal to the direction of an electric field as unlike ends of the domains pull each other. There is a centrosymmetric relation between field and deformation.

Technically valuable qualities of materials are piezoelectric and electrostrictive ceramics, which comprise of haphazardly positioned grains partitioned by grain boundaries. These are cost effective and offer comparable piezoelectric and electrostrictive properties. A piezoelectric material on deformation develops electric dipoles resulting in a potential difference. The reason for developing dipoles is lack of centre of symmetry (anisotropy). This effect is reversible, if a potential difference is applied to such a material it gets deformed. An important point to note is piezoelectric strains are much higher than electrostrictive strains. Piezoelectric crystals being brittle and slightly have larger weight paved way for available piezoceramics which can be sliced into a variety of desirable sizes and shapes and is effortlessly bondable. A piezoelectric system is converted into an intelligent system wherein a sensor and actuator are mutually secured via integrated electronics with intellect capabilities. In order to express piezoelectric properties, the crystal should belong to one of 20 non centrosymmetric crystallographic class.

Piezoelectricity was discovered in 1880 by Jacques Curie and Pierre Curie, however, the term piezoelectricity was coined by W.G. Henkel in 1881. G. Lippman discovered the reverse process that is to say on an application of electric field mechanical stresses could be developed in 1880. Which was experimentally proven by Curie brothers in 1881.

Piezoelectric materials can be natural or synthetic. Natural occurring crystals include topaz, silk, wood etc. While as synthetic are quartz like ceramic polymers and composites. Out of 32 piezoelectric crystal structures, these are further grouped into seven sub-classes as cubic, orthorhombic, hexagonal, trigonal, monoclinic, tetragonal, and triclinic. All of these are elastic in nature, triclinic is anisotropic, orthorhombic is orthotropic material and cubic is anisotropic material. And out of 32, only 20 piezoelectric crystal structures show piezoelectric properties and out of those 20, a mere of 10 such sub-classes are polar that is to say show natural unschooled polarization in the absence of mechanical stress due to existing electric dipole moment amalgamated in their unit cells. The other ten crystal structures are non-polar that is to say polarization transpire only after application of mechanical load.

There are various types of ceramics manufactured by humans with perovskite crystal structure, one of the fundamental crystal lattice structure e.g. Barium titanate $BaTiO_3$, Lead titanate $PbTiO_3$, and Lead Zirconium Titanate $Pb[Zr_xTi_{1-x})O_3]$ has perovskite structure. The chemical structure is ABO_3 type. Wherein A is typically Pb(Ba), the large size metal ion and B ordinarily Ti(Zr), denotes small size metal ion an illustrated in Fig, 1.2. Formation of piezoelectric ceramics require mixing of suitable powders of constituent metal oxides in apt ratio, the mixture is heated to around 800-1000 degree Celsius to obtain consistency. Which is then mixed with an organic binder and shaped into structural elements with requisite forms (rods, discs, plates etc.). These elements are discharged as per particular time and temperature program in the course of which powder particles coalesce into solid mass by means of heating and achieves a dense crystalline structure. The elements upon cooling are sculptured or cropped to desired requirements and electrodes are put across appropriate surfaces. Above Curie temperature which is 150-350 degree Celsius for most of the piezo-ceramics, each ceramic element reveal a cubic symmetry with zero dipole moment, in the phase known as paraelectric phase. At temperatures below Curie temperature, it manifests into a rhombohedral or tetragonal symmetry with a dipole moment, this is the ferroelectric phase. At about 10^6V/m applied electric field to a ferroelectric polycrystalline it passes via Curie temperature, thus spontaneous polarization starts to exist, aligning polarization vectors in more or less uniform direction. This phenomenon is known as polling. After polling a net polarization is established in the ceramic element. At this very point, on application of mechanical stress, the extent of polarization will decrease or increase and will exhibit distinctive piezoelectric behaviour. Piezocomposite materials are an updated version of present piezoceramic. These are classified into two piezopolymers wherein material is dipped in the electrically passive matrix (like PZT immersed in epoxy matrix) and piezocomposites which are in turn made of two piezoceramic materials (like BaTiO3) fibres fortifying a PZT matrix. There exist many crystals having piezoelectric properties, piezoelectric properties get revealed in them as a result of the influence of electromagnetic fields on matter. Piezoelectric materials can be divided into two classes: polar and non-polar piezoelectrics. For example, $Pb_5Ge_3O_3$, PZT, $BaTiO_3$, are polar piezoelectrics, and $Bi_{12}GeO_{20}$, KH_2PO_4, TeO_2, $Bi_{12}SiO_{20}$, are non-polar piezoelectrics. The non-zero dipole moment per unit volume of polar piezoelectric materials (ferroelectrics) in contrast to non-polar piezoelectrics is the only difference between the two and hence polar piezoelectrics possess spontaneous and abrupt polarization. The piezoelectric modules of polar materials are substantial according to the classical theory of piezoelectricity than those of non-polar materials and in fact, the difference that exists between polar and non-polar piezoelectrics.

1.5 Piezoelectric, Pyroelectric and Ferroelectric Materials

All pyroelectric materials are piezoelectric in nature but not all piezoelectric materials are pyroelectric. However ferroelectric forms a subset of the set of pyroelectricity, as they are polar materials in which the direction of the polar axis can be changed on employment of an electric field and as a result they are both pyroelectric and piezoelectric. As many of the largest pyroelectric and piezoelectric effects occur in ferroelectric materials, they have become very important technologically.

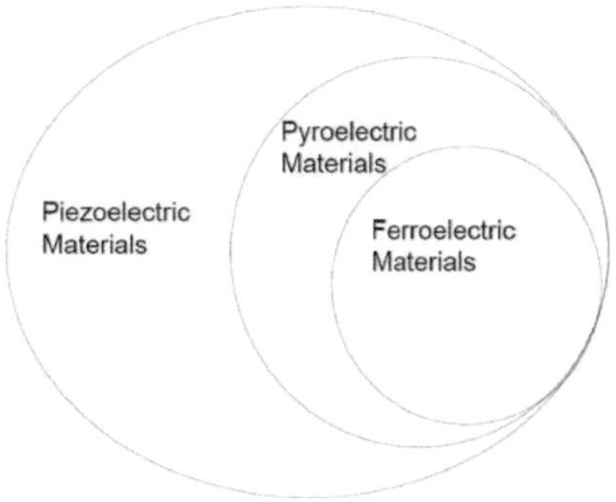

Figure 1.3: Piezo, Pyro and Ferroelectricity

1.6 Lead Zirconium Titanate (PZT)

Lead Zirconium Titanate (PZT) a piezoceramic with a chemical composition [$Pb(Zr_{1-x}Ti_x)O_3$)], is a solid solution of lead zirconate and lead titanate, generally doped with other materials to generate definite properties and is obtained by heating a mixture of oxide powders of lead, zirconium and titanium to about 800 to 1000° C first to get a perovskite PZT powder, which is then mixed with binder and sintered into the desired shape. The resultant unit cell is elongated in one of the directions and exhibit a permanent dipole moment along this axis. Domain is defined as a miniscule region within the ceramic which have same direction of polarization, Since the ceramic, however consist of many randomly oriented domains, it has no net polarisation, Polarisation is created by process of polling where in a material when subjected to high potential difference at very high temperature above its Curie temperature in a particular direction followed by abrupt cooling in order to retain the deformation caused in the crystal. This effect is utilized when a PZT is bonded to a structure inducing positive or negative bending moment. One key factor in polling is the temperature at which polling is performed. As domain wall motion is thermally activated, the polling process is often performed at an elevated temperature to propel a higher degree of domain reorientation, leading to larger polarization and higher piezoelectric coefficient. A higher coefficient of piezoelectricity, high coupling factor and lofty Curie temperature T_c makes material desirable for broader band of piezoelectric applications. In some of the applications piezoelectric materials are required to operate at extremely higher temperatures. A lot needs to be done in order to make suitable piezoelectric materials with higher Curie temperatures (T_c). Since available piezoelectric materials are expected to undergo thermal degradation and faster aging when subjected to soaring temperatures. Degradation of piezo electrical properties due to loss of polarization is called thermally activated aging. Minimization of ageing effect restricts application of materials till half of their curie temperatures.

PZT is a very handy and versatile material. It is chemically dormant and exhibits very high sensitivity. The sensing capability of a PZT patch is utilized in sensing conductance to monitor health of a structure. Key features include low cost, small size, dynamic performance, and fast response, almost unending stability, long range of linearity and high energy conversion efficiency. PZT patches of any shape, size or thickness can be manufactured at a relatively low cost and can be utilised over a broad range of pressures without serious non-linearity. Since it is characterised by a high value of modulus of elasticity. If heated above Curie temperature which varies from 150-350 degree Celsius ferroelectric effect vanishes in the crystals. It can also lose piezoelectricity when subjected to high electric fields above 12 kV/cm reverse to polling direction (depolling) and can lead to permanent change in the dimension of sample. PZT patches are great actuators owing to their high stiffness. Their use vary from single to multi-layered PZT systems e.g. deformable mirrors, mechanical micro positioners, biomorphic actuators. Their brittle character makes them prone to poor conformability to curved surfaces and bending and fluctuation of electrical properties with temperature is another limitation.

1.7 Interactions in Ferroelectric Crystal

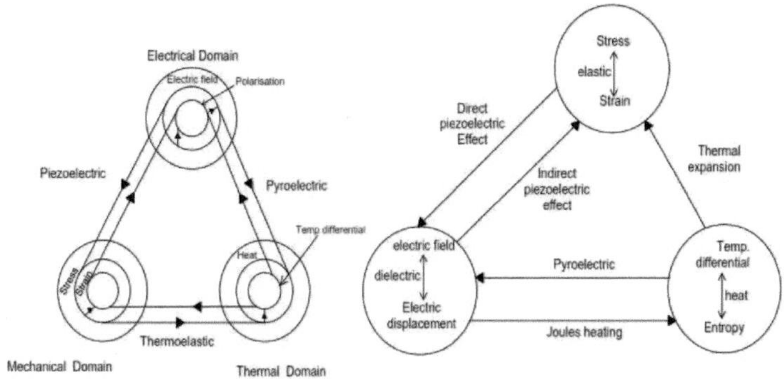

Figure 1.4: Interaction in Electro-Thermo-Elastic Domains

Basically three kinds of physical behaviour prevail in a Ferroelectric crystal. They are piezoelectricity, pyroelectricity and thermo-elasticity, as shown in the above figure. . An electric field in the electric domain generates piezoelectric stress in the mechanical domain, the two being related by the appropriate piezoelectric stress coefficient and is the converse, inverse or indirect piezoelectric effect. Similarly a strain in the mechanical domain generates the electric polarisation in the electrical domain, which in turn produces Electric field, the polarisation and Electric field being associated by dielectric susceptibility. In an akin manner, a temperature differential in the thermal domain and polarisation in the electrical domain are affiliated by pyroelectric constant; the arrowed path from electric field to heat indicates the electro-caloric effect (The generation of an amount of heat δQ on implementation of vector field), generally conveyed as a relationship between temperature differential and electric field. The coefficient of thermal expansion relating temperature differential ΔT with strain, the arrowed path relating stress to heat δQ indicates the thermo-elastic effect. The Constitutive equations link stress and strain in mechanical domain and specific heat relating the quantity of heat δQ and temperature differential in thermal domain. The small arrowed paths in electrical domain between polarisation P and electric field E indicating that polarisation may prevail by existence of electric field and vice versa in ferroelectric material.

The three domains driving each other basically depict the direct or basic effect. However, in every instance there is the minimum of one other way over which the process can occur, of course the coupling coefficients are assumed to be non-zero for that specific process for the crystal or material to be considered. The roundabout indirect effects are called secondary effects. A classic example will be secondary pyroelectric effect owing to piezoelectric action, which is usually many times higher than the primary effect itself. The basic pyroelectric effect is specified by the arrowed path connecting temperature differential with polarization, that is material being polarized on being subjected to temperature differential. Whereas the secondary effect pursues the way temperature differential producing strain which in turn polarizes the material. The logic for this complexity is due to the reason that all pyroelectric crystals are also piezoelectric. Thus a temperature differential to an unconstrained crystal produces a strain (deformation) and this subsequently yields the secondary polarisation superimposed on the basic or primary pyroelectric polarisation.

Summary

A smart material has one or more properties that can be significantly transformed in a constrained or controlled manner by external stimulant/impetus. Study of smart materials has the potential for aspiring future of life long efficient and improved reliability in the use of smart structures. Comprehending and controlling the composition and micro structure of new materials are ultimate objective of research in this field and boom to the manufacture of excellent smart materials. Their application more so rely on the very precise and controllable nature of piezo effect making them immensely invaluable for their area of application. The ability of Ferroelectric materials to transmute loads from electrical, thermal and mechanical domains into each other and back have ensured their use in structures as self-monitoring and self-controlling systems and for smart materials and structures under scrutiny and observations for research. Ferroelectric materials which are subset of pyroelectric materials and in turn subset of piezoelectric materials exhibit both sensing and actuating capabilities simultaneously. They involve interactions in mechanical, electric and thermal domains. The three domains drive each other and represent the primary effect however there is at least one more path connecting the two domains via the third one representing the secondary effect, the secondary effect are to be taken into consideration while modeling the ferroelectric materials.

Chapter -II

MATHEMATICAL MODELING OF AN ELECTRO THERMAL ELASTIC SYSTEM

2.1 Introduction

It requires the fundamental grasp of electrodynamics, continuum mechanics, heat transfer and thermodynamics to model the behaviour of piezoelectric continuum in a smart structure required for formulation of an electro-thermo-elastic system. The electro-thermo-elastic interaction outcomes while modelling the performance of the ferroelectric continuum in a smart structure can be acquired about by establishing in a consistent manner the governing equations, constitutive equations and the boundary conditions. The electrical behaviour of the continuum is modelled in a quasi-static modus operandi taking into account only the laws of electro-statics as the continuum mechanical motion is by various orders smaller than the charge motion and thus enabling to neglect the electro-magnetic phenomena included with the motion of free charges and/or dipoles in insulators.

The forces and moments that act on the polarised system are obtained by applying the concepts of electrostatics of dielectrics amalgamated in the thermo-mechanical modeling of the ferroelectric continuum. The governing equations are obtained by making use of the laws of conservations to the ferroelectric medium, while applying principles of thermodynamics, first and second laws will provide the necessary constitutive relations.

The electro-thermo-elastic problem generates thirty five equations.

- One equation for conservation of mass.
- One equation for conservation of energy.

- Six equations (force and moment equations each three in number) of equilibrium.
- One equation for Gauss law of electrostatics, $D_{i,i} = 0$
- Three equations for electric displacement in terms of electric field and polarisation vectors, $D_j = \varepsilon_0 E_j + P_j$
- Six equations of strain tensor in terms of displacements,

$$\epsilon_{ij} = \frac{1}{2}\left(\frac{\partial u_i}{\partial x_j} + \frac{\partial u_j}{\partial x_i} + \frac{\partial u_k}{\partial x_i}\frac{\partial u_k}{\partial x_j}\right)$$

- Six equations for stress tensor in terms of strain tensor, polarisation vector, electric field vector and temperature, $\sigma_{lm} = C_{lmpq}\epsilon_{pq} - P_l E_m - e_{lmj}E_j - \alpha_{lm}\theta$
- One equation for rate of entropy production equation.
- Three equations for electric displacement vector in terms of strain tensor, electric field vector and scalar temperature, $D_i = e_{imn}\varepsilon_{mn} - b_{in}E_n - \eta_i\theta$
- Three equations for Electric field vector as gradient of scalar electric potential, $E_j = -\phi_{,j}$
- One equation for entropy in terms of strain tensor, electric field vector and temperature,

$$\lambda = \alpha_{mn}\varepsilon_{mn} + \eta_n E_n + C_\theta\theta$$

- Three equations for heat flux in terms of temperature gradient (Fourier Law), $Q_i = -K_{il}\theta_{,l}$

With corresponding nineteen variables in Mechanical domain, ten variables in Electrical and five variables in Thermal domain:

- Mechanical Domain → variables → {one density (ρ_m), nine stress (τ_{ij}) components, three Displacement components (axial displacement u , displacement along depth v and transverse displacement w) and six strain (ε_{ij}) components)}

- Electrical Domain → variables → { Electric field ($\vec{E}$, with $E_x, E_y\ and E_z$ components), Electric displacement ($\vec{D}$, with $D_x, D_y\ and D_z$ components), Polarisation ($\vec{P}, with\ P_x,\ P_y\ and\ P_z$ components) and Electric potential (ϕ, one scalar)}
- Thermal Domain → variables → {Heat flux ($\overrightarrow{Q_\iota}$, $with\ Q_x,\ Q_y\ and\ Q_z$ components), Temperature differential (θ , one scalar) and Entropy (λ , one scalar)}

To derive the above equations in a methodical fashion we start with electrostatics then with principles of conservations and finally with laws of thermodynamics.

2.2 Electrostatics

For a methodical procreation, the electrostatic equations are developed using Coulomb's law for point charges, system of charges, charge continuum, discrete dipoles and dipole continuum.

2.2.1 Coulomb's Law

Point charges q_1 and q* at a distance $\vec{R}$ from each other are as shown in Fig. 2.1. The force exerted on a test charge q* which is at rest and at a span of $\vec{R}$ from the source point charge q_1 is given by Coulombs law as

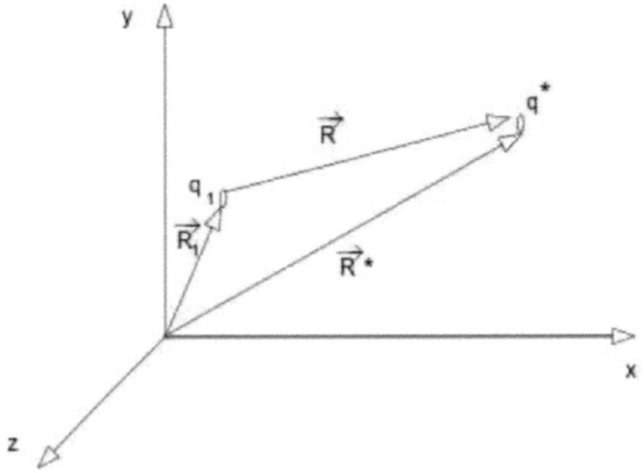

Figure 2.1: Electrostatic Force between Two Charges

$$\vec{F} = \frac{1}{4\pi\varepsilon_0}\frac{q^* q_1}{R^2}\hat{\boldsymbol{R}}$$

$$\vec{F} = \frac{1}{4\pi\varepsilon_0}\frac{q^* q_1}{R^3}\vec{\boldsymbol{R}} \qquad (2.1)$$

The constant ε_0 is the dielectric constant or permittivity of free space or intrinsic capacitance per unit length of free space and its value is ε_0= 8.542×10^{-12} F/m ($C^2/N-m^2$)

2.2.2 Electric Field and Electric Potential

The force F acting on a test charge q* at location (x , y , z) due to a number of charges q_1, q_2, q_3, - - - , q_i,- - - q_n at coordinates (x_1 , y_1 , z_1) , (x_2 , y_2 , z_2),------(x_i , y_i, z_i)----------(x_n , y_n , z_n) is the vector sum of all the forces originating from the separate charges summing up vectorally, the forces due to all charges q_i (i =1,2,- -, n)

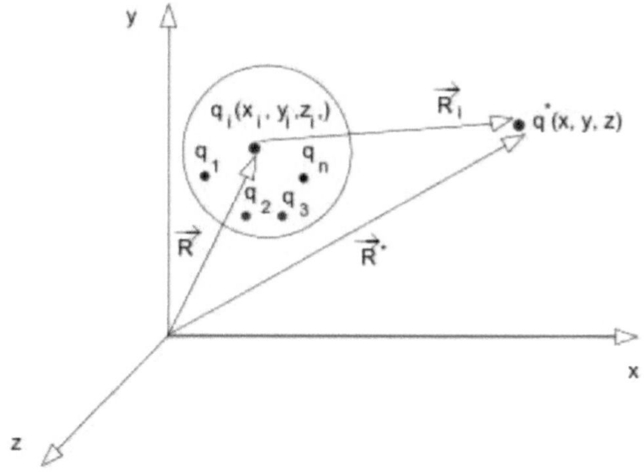

Figure 2.2: Electrostatic Force because of System of Charges at a Point

$$\vec{F} = \sum_i \frac{1}{4\pi\varepsilon_0} \frac{q^* q_i}{R_i^2} \widehat{\boldsymbol{R}}_i$$

$$\vec{F} = \sum_i \frac{1}{4\pi\varepsilon_0} \frac{q^* q_i}{R_i^3} \vec{\boldsymbol{R}}_i$$

(2.2)

Where the vector $\overrightarrow{\boldsymbol{R}_i} = A\overrightarrow{e_x} + B\overrightarrow{e_y} + C\overrightarrow{e_z}$ is from q_i to q^* with magnitude

$R_i = [\,(A)^2 + (B)^2 + (C)^2]^{0.5} for\ A = (x - x_i);\ B = (y - y_i)\ and\ C = (z - z_i)$

(2.3)

The Eq. 2.2, can be rewritten as $\vec{F} = q^* \vec{E}$ (2.4)

Where

$$\vec{E}\ (\boldsymbol{R}) = \frac{1}{4\pi\mathcal{E}_0}\sum_i \frac{q_i}{R_i^2}\hat{R}_i$$

Or

$$\vec{E}\ (\boldsymbol{R}) = \frac{1}{4\pi\mathcal{E}_0}\sum_i \frac{q_i}{R_i^3}\vec{\boldsymbol{R}}_i \tag{2.5}$$

$\vec{E}$ is called the Electric field and is the force that will be exerted by source charges and experienced by test charge per unit of charge if placed at (x , y , z)

Since $R_i = [\,(A)^2 + (B)^2 + (C)^2]^{0.5} with\ A = (x - x_i); B = (y - y_i)\ and\ C = (z - z_i)$

$$\frac{\partial}{\partial x}\left(\frac{1}{R_i}\right) = \frac{-A}{R_i^3}$$

$$\frac{\partial}{\partial y}\left(\frac{1}{R_i}\right) = \frac{-B}{R_i^3}$$

And

$$\frac{\partial}{\partial z}\left(\frac{1}{R_i}\right) = \frac{-C}{R_i^3}$$

That is

$$grad\left(\frac{1}{R_i}\right) = \frac{-R_i}{R_i^3}$$

$$\nabla\left(\frac{1}{R_i}\right) = \frac{-\mathbf{R}_i}{R_i^3} \tag{2.6}$$

Hence we can write $\vec{E} = -\nabla\phi$ (2.7)

Where

$$\phi = \sum_I \frac{1}{4\pi\mathcal{E}_0}\frac{q_i}{R_i} \tag{2.8}$$

ϕ is called the electric potential at point (x, y, z) in volt (N-m/C).

2.2.3 Field and Potential due to Surface and volume Charge Continuum

If instead of being a point charges q_i (i =1, 2,---n) , charge is distributed continuously over some region then for charge continuum

Field and potential caused by the surface charge may be determined by

$$\vec{E}\ (x,y,z) = \frac{1}{4\pi\varepsilon_0}\iint \frac{q(x_1,y_1,z_1)}{R^3}\vec{R}\ dA \quad (2.9)$$

and

$$\phi\ (x,y,z) = \frac{1}{4\pi\varepsilon_0}\iint \frac{q(x_1,y_1,z_1)}{R}dA \quad (2.10)$$

Where $R = [\ (A)^2 + (B)^2 + (C)^2]^{0.5} with\ A = (x - x_i); B = (y - y_i)\ and\ C = (z - z_i)$

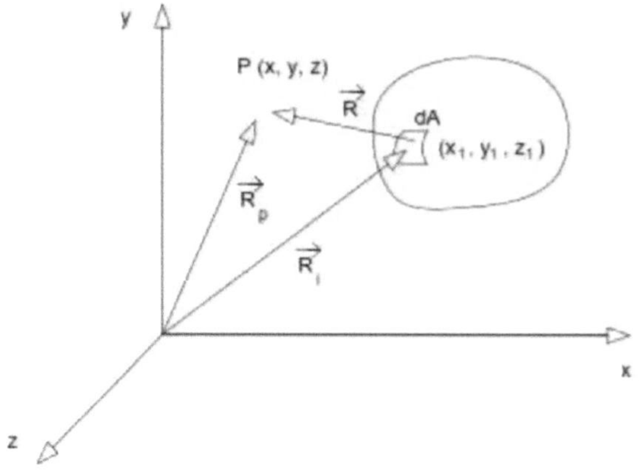

Figure 2.3: Potential and Field due to Surface Charge Continuum

The element dA of the surface has the coordinates (x_1, y_1, z_1) and at that point of the surface the charge per unit area is q

The electric field and potential by a volume charge continuum (for dielectrics) is given as

$$\vec{E}\ (x, y, z) = \frac{1}{4\pi\varepsilon_0} \iiint \frac{\rho(x_1, y_1, z_1)}{R^3} \vec{R}\ dV \tag{2.11}$$

And

$$\phi\ (x, y, z) = \frac{1}{4\pi\varepsilon_0} \iiint \frac{\rho(x_1, y_1, z_1)}{R}\ dV \tag{2.12}$$

Where charge fills a volume with charge density $\rho\ (x_1, y_1, z_1)\ C/m^3$ and the elemental volume is dV.

The Electric Field produced exerts a force $q^*\vec{E}$ on a charge q^*. If this charge moves in the field this force performs work. The infinitesimal amount of work done by the field along the path dS is

$$\delta W = q^*\vec{E}.\overrightarrow{dS} \tag{2.13}$$

When the motion takes place along an arbitrary path from point 1 to point 2 and the vector element of the curve is dS with components dx, dy and dz

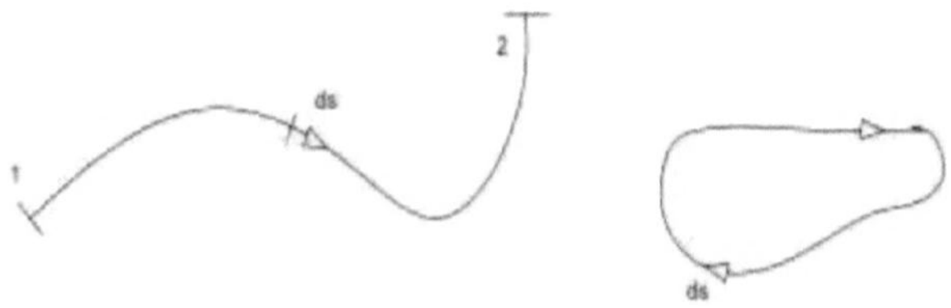

Figure 2.4: Motion of a Charge from Point 1 to Point 2 and Closed Path

That is $\overrightarrow{dS} = dx\,\overrightarrow{e_x} + dy\,\overrightarrow{e_y} + dz\overrightarrow{e_z}$

Then $\delta W = q^* \vec{E}.\overrightarrow{dS}$ gives using Eq. 2.7,

$$\delta W = q^*(-\nabla\phi).(\,dx\,\overrightarrow{e_x} + dy\,\overrightarrow{e_y} + dz\overrightarrow{e_z}\,)$$

$$\delta W = -q^*(\frac{\partial\phi}{\partial x}\vec{e}_x + \frac{\partial\phi}{\partial y}\vec{e}_y + \frac{\partial\phi}{\partial x}\vec{e}_z).(dx\,\overrightarrow{e_x} + dy\,\overrightarrow{e_y} + dz\overrightarrow{e_z}\,)$$

$$\delta W = -q^*(\frac{\partial\phi}{\partial x}dx + \frac{\partial\phi}{\partial y}dy + \frac{\partial\phi}{\partial x}dz)$$

$\delta W = q^* d\phi$ (Potential ϕ is a function of x, y, z)

The total work done in transporting the charge q * from point 1 to point 2 is given as:

$$W_{1-2} = -q^* \int_1^2 d\phi$$

$$W_{1-2} = q^*\,(\phi_1 - \phi_2) \qquad (2.14)$$

This implies that the fall in potential from the starting to the end point multiplied by the magnitude of the charge displaced is equal to the work done. Moreover, it is obvious from Eq. 2.12, that potential is zero at the infinity and it increases as R_i approaches zero. The electric potential can also be defined as the work done in transporting a unit charge from infinity to the location (x, y, z).

The work done per unit displaced charge $\int_1^2 \overrightarrow{E}.\overrightarrow{dS} = \phi_1 - \phi_2$

Choosing the closed path of Integration $\phi_1 = \phi_2$ i.e. $\int_1^2 \overrightarrow{E}.\overrightarrow{dS} = \oint \overrightarrow{E}.\overrightarrow{dS} = 0$

Using Strokes Theorem $\oint \overrightarrow{E}.\overrightarrow{dS} = \iint(\overrightarrow{\nabla} \times \overrightarrow{E}\,).dA = 0$ (2.15)

Which implies $\vec{\nabla} \times \vec{E} = 0$, i.e. $\vec{E}$ is a vector whose curl is zero and hence $\vec{E}$ is equal to the gradient of some scalar (potential). Thus the representation given by Eq. 2.7,

$$\vec{E} = -\vec{\nabla}\phi$$

is validated.

2.2.4 Electric Field in Dielectrics

Dielectrics are materials which offer very high resistance to the passage of direct current that is they are electrically insulating and may be made to exhibit or exhibits an electric dipole structure. Dielectrics are either polar dielectrics that have permanent electric dipole moments with random orientation of molecules in the absenteeism of an electric field. Under the influence of an external field, a torque is born and prompts the molecules to orient with Electric Field, due to random thermal motion however the alignment is not complete. The other type non-polar dielectrics are not blessed with permanent electric dipole moment, but can be incited by placing the nonpolar dielectrics in an externally applied electric field. Thus dielectric are electrically insulating, but susceptible to polarization under the influence and effect of an electric field.

When a non-polar dielectric material is placed in an electric field, the field will incite a tiny dipole moment, pointing in the same direction as the field, in each molecule and for polar dielectric material each permanent dipole will confront a torque tending to line it up along the field orientation.

2.2.5 Potential of an Electric Dipole

An isolated electric dipole consists of two charges of equal magnitude and situated closely. The electric moment of the dipole, $\vec{p}$, which points in the same direction as $\vec{E}$ is defined by $\overrightarrow{p} = q\,\vec{a}$ where $\vec{a}$ is the displacement vector from negative to positive charge.

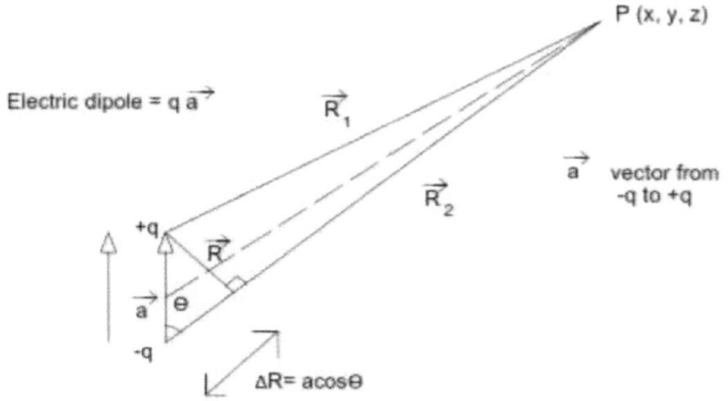

Figure 2.5: An Isolated Electric Dipole

Referring to Fig, 2.5, the potential at point P(x, y, z) because of an isolated dipole can be obtained using Eq. 3.8, and it can be expressed as

$$\phi(x,y,z) = \frac{1}{4\pi\varepsilon_0}\left(\frac{q}{R_1} - \frac{q}{R_2}\right)$$

Where R_1 and R_2 are the spans from $+q$ and $-q$ to the point P(x, y, z)

From the Fig.3.5, $\Delta R = R_2 - R_1 = a\,cos\theta$

$$\phi(x,y,z) = \frac{q}{4\pi\varepsilon_0}\left(\frac{a\,cos\theta}{R_1 R_2}\right)$$

Since $\vec{a} < R_1$ and $\vec{a} < R_2$

And $\overrightarrow{R} \approx R_1 = R_2$

Since $$a\cos\theta = \frac{\vec{a}\,.\,\vec{R}}{R}$$

$$\phi(x,y,z) = \frac{q}{4\pi\varepsilon_0}\left(\frac{\vec{a}\,.\,\vec{R}}{R^3}\right)$$

Also

$$\vec{\nabla}\,\frac{1}{R} = \frac{-\vec{R}}{R^3}$$

Therefore

$$\phi(x,y,z) = \frac{-q\,\vec{a}}{4\pi\varepsilon_0}.\ \vec{\nabla}\,\frac{1}{R}$$

or

$$\phi(x,y,z) = \frac{-\vec{p}}{4\pi\varepsilon_0}.\ \vec{\nabla}\,\frac{1}{R} \qquad (2.16)$$

The formulation is based on a continuum approach, in which the distances are several orders greater than the distance a used in defining the molecular dipole moment. Therefore in the subsequent sections the potential given by Eq. 3.16, is used.

2.2.6 Force and Torque on a Dipole

(a) Force

Consider an isolated dipole placed in an external electric field as shown in Fig. 2.6.

The force on a dipole in a non-uniform field is

$$\vec{F} = \vec{F}_{+} + \vec{F}_{-}$$

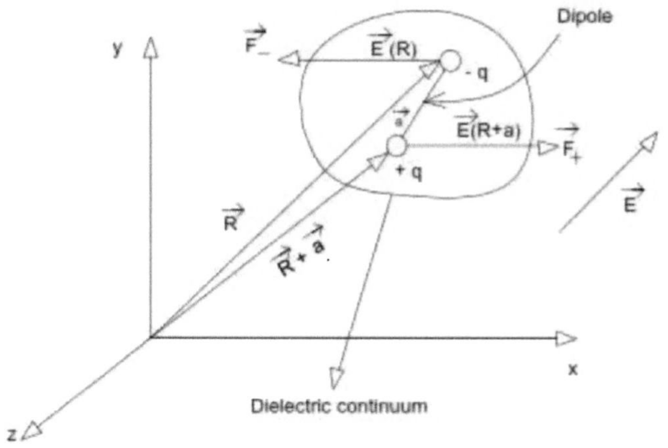

Figure 2.6: Polarised Continuum in an External Electrical Field

Where $\vec{F}_+ \, and \, \vec{F}_-$ are forces on the positive and negative charges of the dipole respectively substituting for $\vec{F}_+ \, and \, \vec{F}_-$ in terms of electric field, the resultant force can be compiled as:

$$\vec{F} = q\,\vec{E}_+ + q\,\vec{E}_- = q\Delta\vec{E}$$

Using Taylor series approximation of $\vec{A}\,(\vec{R} + \vec{r}\,)$ in terms of $\vec{A}\,(\vec{R})$

$$\vec{A}\,(\vec{R}) + \sum_i \{\vec{\nabla}\, A_x . \vec{r}\} i = \vec{A}\,(\vec{R} + \vec{r}\,)$$

Which simplifies to

$$\vec{A}\,(\vec{R}) + (\vec{r}\,.\vec{\nabla})\vec{A} = \vec{A}\,(\vec{R} + \vec{r}\,)$$

Providing Electric field differential as (for $\vec{A} = \vec{E}$ & $\vec{r} = \vec{a}$)

$$\Delta\vec{E} = (\vec{a}\,.\vec{\nabla})\vec{E}$$

Which when substituted into force expression yields:

$$\vec{F} = q\,(\vec{a}\,.\vec{\nabla})\vec{E}$$

In terms of polarisation

The resultant force on a dipole $$\vec{F} = (\vec{p}\,.\vec{\nabla})\vec{E} \tag{2.17}$$

(b) TORQUE

The torque acting on a dipole due to non-uniform electric field on the elemental dipole continuum can be given as:

$$\vec{\tau} = (\vec{a}\ /2) \times \vec{F}_+ + (-\ \vec{a}\ /2) \times \vec{F}_-$$

Substituting for the force acting on positive and negative charges of the dipole due to the electric field, the moment expression becomes:

$$\vec{\tau} = (\vec{a}/2) \times q\vec{E}_+ + (-\vec{a}/2) \times q\vec{E}_-$$

$$\vec{\tau} = (\vec{a}/2) \times q\vec{E} + (-\vec{a}\ /2) \times (-q\,\vec{E}\,)$$

$$\vec{\tau} = \vec{a}q\ \times \vec{E}$$

As $$\vec{E}(\vec{R} + \vec{a}) \approx \vec{E}\,(\vec{R})$$

Therefore $$\vec{\tau} = \vec{p}\ \times \vec{E} \tag{2.18}$$

Polarization $\vec{P}$ is defined as dipole moment per unit volume, therefore:

Force on a dipole continuum per unit volume (force density $\vec{F}_p$)

$$\vec{F}_p = (\vec{P}\,.\vec{\nabla})\vec{E} = P_l E_{i,l} \vec{e}_i \tag{2.19}$$

Moment on a dipole continuum per unit volume (moment density $\vec{\tau}_p$)

$$\vec{\tau}_p = \vec{P}\ \times \vec{E} = \epsilon_{ijk} \vec{e}_i P_j E_k \tag{2.20}$$

2.2.7 The Field of a polarized Dipole Continuum

Consider a volume which encloses matter and it contains positive and negative point charges q_1 , q_2, q_3, $- - -$ q_n which may be of different sizes but such that $\sum q_i = 0$

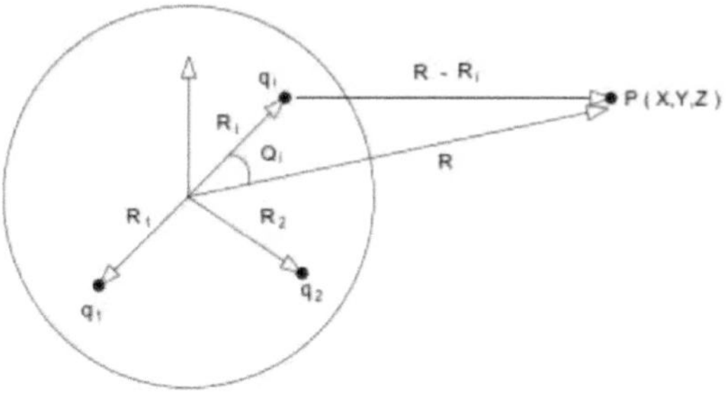

Figure 2.7: Volume Charge Continuum

The potential at site P (x, y, z) can be written as:

$$\phi(x,y,z) = \frac{1}{4\pi \mathcal{E}_0} \sum_i {q_i}/{|R - R_i|}$$

Now Since $$|R - R_i| = \sqrt{(R^2 + R_i^2 - 2RR_I \cos\theta_I)}$$

As $$R >> R_i$$

$$[|R - R_i|]^{-1} = \frac{1}{R\,(1 + \frac{R_i^2}{R^2} - 2\frac{R_i cos\theta_i}{R})^{0.5}}$$

$$[|R - R_i|]^{-1} = R^{-1}\,(1 - 2\,(R_i \cos\theta_i)/R \;+\; R_i^2/R^2\;)^{-0.5}$$

$$[|R - R_i|]^{-1} = R^{-1}\,(\,1 + \frac{R_i}{R}\cos\theta_i + - - -)$$

Therefore

$$\phi(x,y,z) = \frac{1}{4\pi \mathcal{E}_0} \sum_i q_i\,\frac{1}{R}\,(\,1 + \frac{R_i}{R} cos\,\theta_i + - - -)$$

$$\phi(x,y,z) = \frac{1}{4\pi\varepsilon_0}\frac{1}{R}\sum_i q_i + \frac{1}{4\pi\varepsilon_0}\sum_i \frac{q_i\, R_i\, \cos\theta_i}{R^2}$$

As

$$\sum_i q_i = 0$$

Then

$$\phi(x,y,z) = \frac{1}{4\pi\varepsilon_0}\sum_i \frac{q_i\, \boldsymbol{R}_i \,.\, \boldsymbol{R}}{R^3}$$

The dipole moment of the region is defined as

$$\vec{p} = \sum_i q_i\, R_i$$

and using Eq. 3.6,

$$\boldsymbol{\nabla}\left(\frac{\mathbf{1}}{\boldsymbol{R}}\right) = \frac{-\boldsymbol{R}}{R^3}$$

Thus

$$\phi(x,y,z) = -\frac{1}{4\pi\varepsilon_0}\vec{p}.\,\vec{\nabla}\left(\frac{1}{R}\right) \tag{2.21}$$

Dividing the polarised region into two groups, one containing only positive charges and the other only negative.

$$\sum q_i = \sum q_+ + \sum q_-$$

Where

$$\sum q_+ = \sum q_-$$

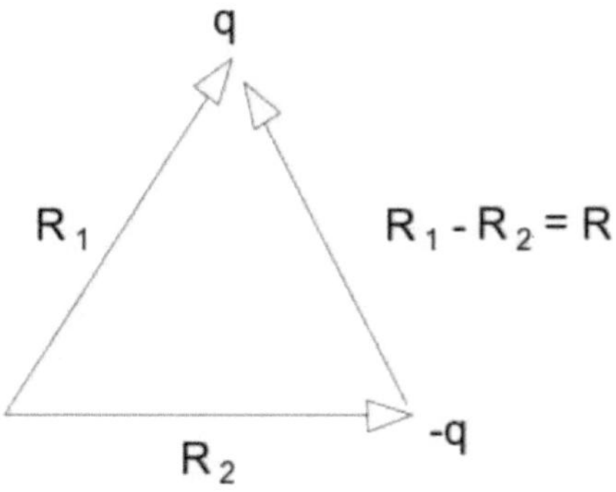

Figure 2.8: Charge Centres of Dipole Continuum

And Position Vectors $\boldsymbol{R_1}$ and $\boldsymbol{R_2}$ for the electric centre of the positive and negative charges

$$qR_1 = \sum q^+ R^+$$

And

$$qR_2 = -\sum q^- R^-$$

then from figure

$$\sum q_i R_i = \sum q^+ R^+ + \sum q^- R^-$$

$$\sum q_i R_i = qR_1 - qR_2 = q(R_1 - R_2)$$

The moment of the region is then

$$\vec{p} = \sum q_i R_i = q\vec{R}$$

Where q is the sum of the Positive charges and the vector **R** is drawn from the electric centre of negative charges to that of the positive charges.

Consider a piece of dielectric material bounded by the surface S

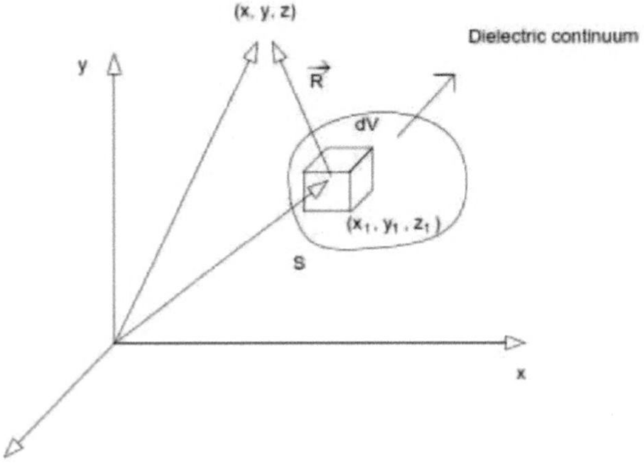

Figure 2.9: Field at a Point due to Dielectric Continuum

If the material is polarized, the potential at the point (x, y, z) within or outside the region is as per Eq. 2.21, and is equal to

$$\phi(x,y,z) = \frac{1}{4\pi\varepsilon_0}\iiint \vec{p}\,.\vec{\nabla}_x \left(\frac{1}{R}\right) dV$$

Where $\vec{R} = A\vec{e_x} + B\vec{e_y} + C\vec{e_z}$ with $R = [\,(A)^2 + (B)^2 + (C)^2]^{0.5}$

$for\ A = (x - x_i); B = (y - y_i)\ and\ C = (z - z_i)$

Subscript x of the gradient indicates that it should be formed in respect to the field coordinates

Since

$$\frac{\partial R}{\partial x} = \frac{A}{R} = -\frac{\partial R}{\partial x_1}$$

$$\frac{\partial R}{\partial y} = \frac{B}{R} = -\frac{\partial R}{\partial y_1}$$

$$\frac{\partial R}{\partial z} = \frac{C}{R} = -\frac{\partial R}{\partial z_1}$$

That is

$$\vec{\nabla}_x\left(\frac{1}{R}\right) = -\vec{\nabla}_{x_1}\left(\frac{1}{R}\right)$$

And thus

$$\phi(x,y,z) = \frac{1}{4\pi\mathcal{E}_0}\iiint \vec{p}\,.\vec{\nabla}_{x_1}\left(\frac{1}{R}\right)\, dV$$

Since $\quad \boldsymbol{\nabla}.(\varphi\boldsymbol{a}) = \varphi\,\boldsymbol{\nabla}.\boldsymbol{a} + \boldsymbol{a}.\boldsymbol{\nabla}\varphi$

If $\quad \varphi = \frac{1}{R}$ and $\boldsymbol{a} = \boldsymbol{P}$

$$\boldsymbol{\nabla}.\left(\frac{P}{R}\right) = \frac{1}{R}\boldsymbol{\nabla}.\boldsymbol{P} + \boldsymbol{P}.\boldsymbol{\nabla}\frac{1}{R}$$

Or $\quad \boldsymbol{P}.\boldsymbol{\nabla}\frac{1}{R} = -\frac{1}{R}\boldsymbol{\nabla}.\boldsymbol{P} + \boldsymbol{\nabla}.\left(\frac{P}{R}\right)$

Therefore

$$\phi(x,y,z) = \frac{1}{4\pi\mathcal{E}_0}\iiint -\frac{div\,\vec{P}}{R}dV + \frac{1}{4\pi\mathcal{E}_0}\iiint div\,(\frac{\vec{P}}{R})dV$$

Using divergence theorem for the second integral, we have

$$\phi(x,y,z) = \frac{1}{4\pi\mathcal{E}_0}\iiint -\frac{div\,\vec{P}}{R}dV + \frac{1}{4\pi\mathcal{E}_0}\iint \frac{\vec{n}\,.\vec{P}}{R}\,dA \qquad (2.22)$$

Here $\vec{n}$ is identity vector perpendicular to the dielectric continuum and source coordinates (x_1, y_1, z_1) are used to perform differentiation.

Comparing first term with Eq. 2.12,

$$\rho^b(x_1,y_1,z_1) = -\vec{\nabla}\,.\vec{P} \qquad (2.23)$$

That is potential of a volume charge

Comparing the second term with Eq, 2.10,

$$q^b(x_1,y_1,\,z_1) = \vec{n}.\vec{P} \qquad (2.24)$$

That is potential of a surface charge density of a polarized dielectric continuum is identical to that generated by surface charge density $q^b = \vec{n}.\vec{P}$ added by that yielded by the volume charge density, $\rho^b = -\vec{\nabla}.\vec{P}$. The charge distribution due to polarization is bound charge distribution as these charges are not free to move as free charges move in a conductor under potential difference.

Suppose in addition to the field owing to polarisation of the medium. We have field because of free charge ρ^f, free ions embedded in dielectric material

Within the dielectric, then the total charge density

$$\rho = \rho^b + \rho^f$$

As per gauss law

$$\varepsilon_0 \vec{\nabla}.\vec{E} = \rho$$

$$\varepsilon_0 \vec{\nabla}.\vec{E} = \rho^b + \rho^f$$

$$\varepsilon_0 \vec{\nabla}.\vec{E} = -\vec{\nabla}.\vec{P} + \rho^f$$

$$\varepsilon_0 \vec{\nabla}.\vec{E} + \vec{\nabla}.\vec{P} = \rho^f$$

$$\vec{\nabla}.(\varepsilon_0 \vec{E} + \vec{P}) = \rho^f$$

The expression $\varepsilon_0 \vec{E} + \vec{P}$ is known as electric displacement $\vec{D}$

Therefore

$$\vec{\nabla}.\vec{D} = \rho^f \qquad (2.25)$$

And

$$\varepsilon_0 \vec{E} + \vec{P} = \vec{D} \qquad (2.26)$$

The electrostatic equation for a dielectric continuum in absence of free charges is

$$\vec{\nabla}.\vec{D} = 0 \qquad (2.27)$$

2.2.8 Linear Dielectrics

For many substances, the polarization is proportional to the field, provided external electric fields is not too strong.

Hence one can write:

$$\overrightarrow{P} = \varepsilon_0 \, \chi_e \, \overline{E} \qquad (2.28)$$

Where χ_e is labelled for medium as its electric susceptibility. Electric displacement is written in terms of electric field as:

$$\overrightarrow{D} = \varepsilon_0 \, \overrightarrow{E} + \overline{P}$$

$$\overrightarrow{D} = \varepsilon_0 \, \overrightarrow{E} + \varepsilon_0 \, \chi_e \, \overline{E}$$

$$\overrightarrow{D} = \varepsilon_0 \, (1 + \chi_e)\overline{E}$$

$$\overrightarrow{D} = \varepsilon \, \overrightarrow{E} \qquad (2.29)$$

Where $\varepsilon = \varepsilon_0 \, (1 + \chi_e)$ is labelled for materials as their permittivity.

2.2.9 Continuity Conditions between Two Dielectric Materials

Applying electrostatic relations to a pill-box at the interface, to obtain the continuity conditions of electric displacement $\overrightarrow{D}$, electric field $\overrightarrow{E}$ and electric potential ϕ at the interface between two different dielectric materials

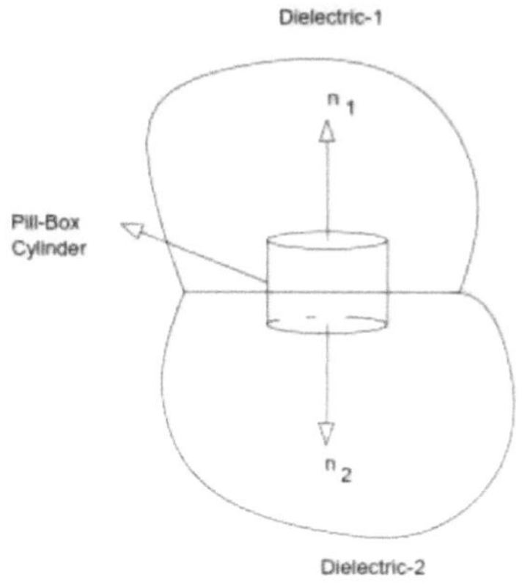

Figure 2.10: Pill-Box at the Interface of two Dielectrics

(a) Electric Displacement

Applying Eq. 3.25, to a pill-box region at the interface as shown in figure and integrating over the volume of the pill-box one gets:

$$\iiint div\,\vec{D}\,dV = \iiint \rho^f\,d\,V + \iint q^f\,dA$$

Applying Gauss divergence theorem, one obtains:

$$\iint \vec{n}\,.\vec{D}\,dA = \iiint \rho^f\,d\,V + \iint q^f\,dA$$

In the limit, reducing the thickness of the pill-box, the sides of the pill-box rectangle normal to the interface shrink to two points. As a consequence, the contribution of the volume charge density ρ^f tends to zero. Hence

$$\iint \vec{n}\,.\vec{D}\,dA = \iint q^f\,dA$$

Splitting the surface integral into two components above equation becomes

$$\iint \vec{n}_1\,.\,\vec{D}_1\,dA + \iint \vec{n}_2\,.\vec{D}_2\,dA = \iint q^f\,dA$$

Where $\vec{D}_1 and\ \vec{D}_2$ represent the electric displacement in the two dielectric materials. Rearranging the integrals:

$$\iint (\vec{n}_1\,.\,\vec{D}_1\ + \vec{n}_2\,.\vec{D}_2\ -\ q^f\,)\,dA\ = 0$$

Since this equality is valid for any arbitrary surface at the interface, the integrant must be identically equal to zero. Therefore one can write the interface conditions between the two dielectrics as:

$$\vec{n}_1\,.\,\vec{D}_1\ + \vec{n}_2\,.\vec{D}_2 = q^f \qquad (2.30)$$

If the free charge density is zero then the interface condition becomes:

$$\vec{n}_1\,.\,\vec{D}_1 = -\,\vec{n}_2\,.\vec{D}_2$$

Since the two unit normal are equal and opposite one can write:

$$\vec{D}_1^n = \vec{D}_1^n \tag{2.31}$$

Where the superscript represents the normal component of the electric displacement in the two dielectrics at the interface. Above equation implies the continuity of normal component of electric displacement at the interface between two dielectric materials.

(b) Electric Field

Consider a closed contour at the interface as shown in figure

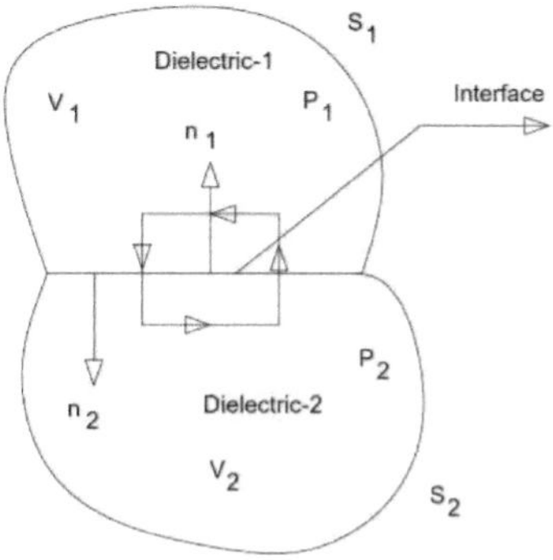

Figure 2.11: Interface Condition between Two Dielectrics

In the limit of reducing the contour lines normal to the interface, the contour integral given by Eq. 2.15, can be put as:

$$\int \vec{E}_1 . \overrightarrow{dR_1} + \int \vec{E}_2 . \overrightarrow{dR_2} = 0$$

Where the subscript 1 and 2 represent the two materials.

In the limit of reducing the line segments in the integrals,

$$\vec{E}_1^t = \vec{E}_1^t \tag{2.32}$$

Where the superscript t represents the tangential component. Above equation implies that the tangential constituent of the electric field is continuous at the interface between the two dielectric surfaces in contact.

(c) Electric Potential

Previous Eq. 3.32 can be written in terms of tangential derivative of potentials in the two dielectrics as:

$$\frac{\partial \phi_1}{\partial t} = \frac{\partial \phi_2}{\partial t}$$

Where t represents the tangential line element on the interface, ϕ_1 and ϕ_2 correspond to the potential at a point on the interface between the two dielectrics. From the above equation it can be stated that:

$$\phi_1(t, n) = \phi_2(t, n) + f\,(n)\, + C$$

Where f is a function of normal coordinate to the interface and C is a constant. If there is a jump in the values of the potentials ϕ_1 and ϕ_2 at the interface, it implies an infinite electric field in the normal direction at the interface which is physically not possible. Therefore f (n) must be equal to zero at the interface. Hence the continuity condition for the potential at the interface of the two dielectrics can be written as

$$\phi_1 = \phi_2 \tag{2.33}$$

2.2.10 Energy of a Charged-Continuum (Polarised-Medium)

In a non-polarised medium, the energy of a charge continuum in terms of potential ϕ and volume charge density ρ can be given as

$$W = \frac{1}{2}\iiint \rho\, \phi\, dV$$

Substituting for charge density ρ from gauss law $div\,\vec{D} = \rho$ or $div\,\varepsilon_0\vec{E} = \rho$,

$$W = \frac{\varepsilon_0}{2}\iiint \phi\,(\,div\,\vec{E}\,)dV$$

Using $$\nabla.(\varphi \boldsymbol{a}) = \varphi \nabla.\boldsymbol{a} + \boldsymbol{a}.\nabla\varphi$$

With $$\varphi = \phi \text{ and } \boldsymbol{a} = \vec{\boldsymbol{E}}$$

$$div(\phi \vec{\boldsymbol{E}}) = \phi(div \vec{\boldsymbol{E}}) + \vec{\boldsymbol{E}} .grad\phi$$

$$W = \frac{\varepsilon_0}{2}\iiint \{div(\phi \vec{\boldsymbol{E}}) - \vec{\boldsymbol{E}}.grad\phi\} \, dV$$

Now $\vec{E} = -grad\,\phi$, and utilising divergence theorem, the integral can be put as:

$$W = \frac{\varepsilon_0}{2}\{\iiint E^2 \, dV + \iint \phi \vec{E} .\vec{dA}\}$$

It may be noted that ϕ decreases as $\frac{1}{R}$, $\vec{E}$ decreases as $\frac{1}{R^2}$ and area increases as R^2.Therefore the second integral vanishes for large values of R. Therefore energy of a charged-continuum can be written as:

$$W = \frac{\varepsilon_0}{2}\iiint E^2 \, dV$$

In the case of polarised medium, there are both bound and free charge distribution. The energy associated with bringing the free charge in to the polarised medium can be written as:

$$W = \frac{1}{2}\iiint \rho^f \, \phi \, dV$$

Substituting for ρ^f

$$W = \frac{1}{2}\iiint (div \vec{D}) \phi \, dV$$

Using the vector identity $div(\phi\vec{D}) = \phi\,(div\,\vec{D}) + \vec{D}.grad(\phi)$, the above integral can be written as:

$$W = \frac{1}{2}\iiint \{(div(\phi\vec{D}) - \vec{D}.grad(\phi)\} \, dV$$

Again $\overrightarrow{E} = -grad\ \phi$, and employing divergence theorem, the integral can be put down as:

$$W = \frac{1}{2}\{\iiint \overrightarrow{D}\ .\overrightarrow{E}\,dV + \iint \phi\overrightarrow{D}\ .\overrightarrow{dA}$$

By the same argument given above, the second integral vanishes for large values of R. Therefore energy of a polarised dielectric continuum can be written as:

$$W = \frac{1}{2}\iiint \overrightarrow{D}\ .\overrightarrow{E}\,dV \qquad (2.34)$$

2.3 Thermo-Elasticity of a polarized continuum

To obtain with respect to the deformed configuration the governing differential equations and the suitable constitutive equations for the polarized continuum system. The system under consideration is subjected to the laws of thermodynamics and conservation laws. The Conservation of momentum laws provide the necessary governing differential equations. Whereas the thermal equilibrium equation and constitutive relations correspond to first and second laws of thermodynamics applied to the system under consideration.

2.3.1 Definitions of Thermodynamic System

A thermodynamic system is real or imaginary portion of the matter separated from its surroundings and on which the attention is focused for investigation of various physical phenomena taking place in that portion of the matter. Anything outside the system, which is responsible for changing the behaviour of the system is said to be its surroundings. There are several measurable macroscopic parameters and observable characteristics of the system, e.g. Energy, density, pressure and temperature of the thermodynamic system which undergoes changes during the action of the surroundings on the system, they are known as properties of the system. When the properties of the system change under an external influence, the system is said to undergo a change of state or it has undergone a thermodynamic process. In solids it is generally assumed that the chemical composition of the material does not undergo any change during a thermodynamic process. When the surroundings are assumed to have no influence on the system, then such a system is called an isolated system. Mechanical equilibrium of the system corresponds to the balancing of forces and moments on the system according to Newton's laws of motion. The system is in thermal equilibrium when each part of the system is at the identical temperature and all parts of system have temperature same that of surroundings and as a result there are no changes in the properties of the system. Thermodynamic equilibrium correspond to Chemical equilibrium, Mechanical equilibrium and Thermal equilibrium.

2.3.2 Description of Deformation of Body

(a) Material Description or Lagrangian Coordinates

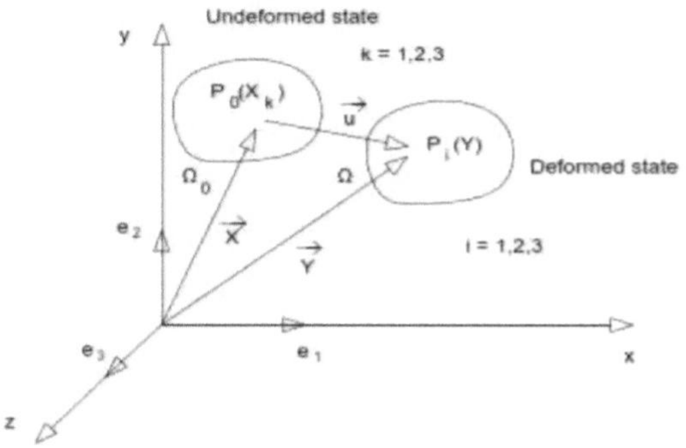

Figure 2.12: Representation of deformational kinematics

In Lagrangian coordinate system (initial condition) at time (t) = 0 the vector representing position of point P_0 can be listed as

$$\vec{R}_L = X_1 \vec{e}_1 + X_2 \vec{e}_2 + X_3 \vec{e}_3 = X_L \vec{e}_L$$

The undeformed position in conational format is penned as; $\vec{X}_L \equiv (X_1, X_2, X_3)$. (The subscript L in upper case denotes the undeformed coordinates, L=1, 2, 3). If the undeformed state is taken as the reference system for describing the position of any material point during elastic deformation, then the reference coordinate system is called material, Lagrangian, referential or initial coordinate system.

(b) Spatial Description or Eulerian Coordinates

When elastic deformations occur, the position of the referential point P_0 at present instant of time t is represented by P_i. In the Eulerian coordinates (current condition) the vector representing the position of the point at t can be listed as:

$$\vec{R}_E = y_1 \vec{e}_1 + y_2\vec{e}_2 + y_3\vec{e}_3 = y_i\vec{e}_i$$

In conational format the deformed position is penned as $\vec{y}_i \equiv (y_1, y_2, y_3)$. (The subscript i in lower case denotes the deformed coordinates, i=1, 2, 3). If the deformed state is taken as the reference system for describing the position of any material point during elastic deformation, then the reference coordinate system is called Eulerian, instant or current coordinate system.

The elastic deformation vector $\vec{u}$ couples the position vectors in the undeformed and deformed state.

$$\vec{u} = \vec{R}_E - \vec{R}_L$$

Revealing the vectors in component form gives:

$$y_1 \vec{e}_1 + y_2\vec{e}_2 + y_3\vec{e}_3 = (X_1 + \vec{u}_1)\vec{e}_1 + (X_2 + \vec{u}_2)\vec{e}_2 + (X_3 + \vec{u}_3)\vec{e}_3$$

The deformation vector components u_k (k= 1, 2, 3) are functions of undeformed state coordinates X_L. With respect to undeformed coordinates the gradients of deformation are written as:

$$y_{i,L} = \frac{\partial y_i}{\partial X_L} \qquad i, L = 1, 2, 3 \tag{2.35}$$

The position vector in undeformed state for Eulerian description is;

$$\vec{R}_L = \vec{R}_E - \vec{u}$$

In this case the components u_k (k = 1, 2, 3) of deformation vector are functions of deformed coordinates, y_j (j = 1, 2, 3). With respect to deformed coordinates the gradients of deformation are written as:

$$X_{L,j} = \frac{dX_L}{dy_j} \qquad L, j = 1,2,3 \tag{2.36}$$

2.3.3 Stress

In an undeformed body, the molecular arrangement correspond to a state of equilibrium that is considering a portion of the body, the vector summation of forces is zero on that portion.

For deformable body subjected to mechanical, thermal or electric loads, the molecular arrangement gets changed and it halts to be in its initial equilibrium state. Forces emerge within the body which resist deformation and under external loads equilibrium is established at another equilibrium state. The intensity of these internal resistive forces which emerge for deformable body subjected to loadings are called stresses. The intensity is calculated per unit area at a point. The phenomenon of generation of restoring forces after deformation is different in polarised and non-polarised mediums. In non-polarised medium the restoring forces are due to molecular interactions whereas in polarised medium these forces are generated by both molecular interaction and by macroscopic induced electric field due to polarisation.

Consider a body deformed because of mechanical, thermal and electrical loads as a result of which internal forces will be induced on any of its elemental volume ΔV.

Assuming F to be force density then aggregate force on volume element dV of the body is $\iiint F\,dV$. This force is the resultant of all the forces exerted on a given part of the body by portions neighbouring it.

The forces with which several parts of the above portion influence on each other gave a zero resultant force as they nullify by third law of motion. The force acting on a given portion then act only on the surface of that portion.

All the three components $\int F_i\,dV$ of the resultant of all the interior forces can be altered and transformed into a surface integral each for any portion of the body, if F_i is a divergence of tensor of order 2.

$$\iiint F_i\,dV = \iiint \frac{\partial \sigma_{il}}{\partial x_l} dV = \iint \sigma_{il}\,dA_l \tag{2.37}$$

Where dA_l are the elemental surface vector constituents directed along outward normal.

(a) Stress in Non-polarised Medium

The arrangement of molecules in an undeformed non-polarised medium is in a thermal equilibrium state. In thermal equilibrium all the elemental parts of the body are also in mechanical equilibrium. Therefore mechanical equilibrium of the part implies that the resultant of all the forces acting on each elemental volume is zero. When the body undergoes a mechanical deformation, the original thermal equilibrium state of the molecules changes due to heat of deformation (Whereas if one considers a quasi-static deformation, the heat of deformation is neglected. In the present formulation the heat generated during mechanical deformation is not considered. However the heat due to thermal source is considered). In the deformed state the internal forces are generated which try to bring the body to the original equilibrium state. These restoring forces are stresses in non-polarised medium.

During deformation intermolecular forces are generated. These molecular forces act in molecular range of distances of the order of 10^{-10} m ~ 10^{-12} m. Continuum mechanics is a macroscopic theory and hence the distances over which the internal stresses act are of macroscopic dimensions, i.e. of the order of 10^{-8} m. In continuum mechanics approach these intermolecular forces are replaced by near action forces which act in the range of macroscopic distances. These near action forces are responsible for the generation of internal stresses.

The second order tensor σ_{ij} is stress tensor because of near action forces in a non-polarised medium. In polarised medium there is an additional stress tensor because of the interaction of polarisation and field. This additional stress is known as Maxwell stress.

(b) Maxwell's Stress

Consider an elemental volume of polarised medium in the deformed state. For the presence of free charges in the polarised medium, then the force acting per unit volume has two parts [2]. These two parts of the force acting on an elemental volume are given by:

Force per unit volume because of polarisation of the medium Eq. 2.19, $(\vec{P}.\vec{\nabla})\vec{E} = P_l E_{i,l}\,\vec{e}_i$

Force per unit volume because of the existence of free charges, if any Eq. 2.25,

$$\rho^f \vec{E} = (\vec{\nabla}.\vec{D})\,\vec{E} = D_i E_{j,i}\vec{e}_j$$

Therefore one can write the total force experienced by an elemental volume δV of the polarised medium as:

$$\iiint \vec{f}\ dV = \iiint \{\,(\vec{P}.\vec{\nabla})\vec{E} + (\vec{\nabla}.\vec{D})\,\vec{E}\,\}\,dV \qquad (2.38)$$

$$\iiint f_j \vec{e}_j dV = \iiint (P_l E_{j,l} \vec{e}_j + D_i E_{j,i} \vec{e}_j) dV$$

Using $P_l = D_l - \varepsilon_0 E_l$ we get $f_j = \left(D_l E_j - \frac{1}{2}\varepsilon_0 \delta_{jl} E^2\right)_{,l}$

Suppose $\vec{f} = f_1\vec{e_1} + f_2\vec{e_2} + f_3\vec{e_3} = f_i\vec{e}_i$ where

$$f_2 = \left(\frac{\partial D_1}{\partial x} + \frac{\partial D_2}{\partial y} + \frac{\partial D_3}{\partial z}\right) E_2 + (P_1 \frac{\partial}{\partial x} + P_2 \frac{\partial}{\partial y} + P_3 \frac{\partial}{\partial z}) E_2$$

Or

$$f_2 = \left((\frac{\partial (E_2 D_1)}{\partial x} - D_1 \frac{\partial E_2}{\partial x}) + (\frac{\partial (E_2 D_2)}{\partial y} - D_2 \frac{\partial E_2}{\partial x}) + (\frac{\partial (E_2 D_3)}{\partial z} - D_3 \frac{\partial E_2}{\partial x}) \right) + (P_1 \frac{\partial}{\partial x}$$

$$+ P_2 \frac{\partial}{\partial y} + P_3 \frac{\partial}{\partial z}) E_2$$

Now $\vec{P} = \vec{D} - \varepsilon_0 \vec{E}$

$$f_2 = \left(\frac{\partial (E_2 D_1)}{\partial x} + \frac{\partial (E_2 D_2)}{\partial y} + \frac{\partial (E_2 D_3)}{\partial z}\right) - \varepsilon_0 (E_1 \frac{\partial E_2}{\partial x} + E_2 \frac{\partial E_2}{\partial y} + E_3 \frac{\partial E_2}{\partial z}) E_2$$

Or

$$f_2 = \frac{\partial}{\partial x}(E_2 D_1) + \frac{\partial}{\partial y}\left(E_2 D_2 - \frac{1}{2}\varepsilon_0 E^2\right) + \frac{\partial (E_2 D_3)}{\partial z}$$

Where $E^2 = E_1^2 + E_2^2 + E_3^2$

And similarly f_1 and f_3 can be written as:

$$f_1 = \frac{\partial}{\partial x}(E_1 D_1 - \frac{1}{2}\varepsilon_0 E^2) + \frac{\partial}{\partial y}(E_1 D_2) + \frac{\partial (E_1 D_3)}{\partial z}$$

$$f_3 = \frac{\partial}{\partial x}(E_3 D_1) + \frac{\partial}{\partial y}(E_3 D_2) + \frac{\partial}{\partial z}(E_2 D_3 - \frac{1}{2}\varepsilon_0 E^2)$$

Substituting f_1, f_2 and f_3 in Eq.2.37, for stresses in a polarised medium

$\iiint f_i \, dV = \iiint \frac{\partial \sigma_{il}^M}{\partial x_l} dV = \iint \sigma_{il}^M \, dA_l$ i = 1, 2, 3

The components of σ_{il}^{M}, the Maxwell stress tensor are given as:

$$\sigma_{il}^{M} = D_l E_i - \frac{1}{2}\varepsilon_0\, \delta_{il}\, E^2 \qquad (2.39)$$

2.4 Equilibrium Equations

Applying principals of mass conservation and momentum conversations to an infinitesimal element of body, the resultant equations are the required equilibrium equations.

(a) Conservation of mass

Consider an elemental volume δV enclosed by S. The mass remains constant during deformation as per law of conservation of mass inside that bounded surface. Therefore the time rate change of mass has to be zero:

$$D_t \iiint \rho_m \, dV$$

ρ_m is density in Eularian coordinate system. Using transport theorem [8], the integral can be modified as:

$$\iiint \left(\frac{d\rho_m}{dt} + \rho_m div\, \vec{v}\right) dV = 0$$

Where $\vec{v}$ is the local velocity, as this equality is applicable for any arbitrary volume, the conservation of mass equation becomes:

$$\frac{d\rho_m}{dt} + \rho_m div\, \vec{v} = 0 \qquad (2.40)$$

(b) Conservation of Linear Momentum

Let Ω_0 be the initial undeformed original configuration and Ω be the present deformed domain or state of the body as illustrated in Fig. 2.11. For the randomly chosen element of volume δV, the surrounding forces and moments per unit volume experienced by it are:

Gravity force: $\vec{B} = B_l \vec{e}_l$

Near action force: $\vec{F} = F_m \vec{e}_m$

Force because of polarisation of the medium: $\vec{F}_p = (\vec{P}.\vec{\nabla})\vec{E} = P_l E_{i,l}\,\vec{e}_i$

Force due to the existence of free charges in the polarised medium, if any: $(\vec{\nabla}.\vec{D})\,\vec{E} = D_i E_{j,i} \vec{e}_j$

Moment because of polarisation of the medium: $-(\vec{E} \times \vec{P}) = \epsilon_{ijk} \vec{e}_i P_j E_k$

Suppose force due to existence of free charge is zero i.e., absence of any free charges in the continuum, then linear momentum principle in Eulerian coordinate system is:

$$\iiint B_i \vec{e}_i dV + \iiint F_i \vec{e}_i\, dV + \iiint P_l E_{i,l}\,\vec{e}_i\, dV \;=\; D_t \iiint \rho_m v_i \vec{e}_i\, dV$$

Where integration is performed over an arbitrary portion of infinitesimal volume in the current deformed state of the domain Ω, let elemental volume has an absolute velocity $\vec{v}$ and ρ_m is the density in the deformed state.

Using transport theorem for the time integral term, the above integral can be modified as:

$$\iiint \vec{F}\, dV + \iiint [\vec{B} + (\vec{P}.\vec{\nabla})\vec{E}]dV = \iiint [\frac{d(\rho_m \vec{v})}{dt} + \rho_m \vec{v}\ div\ \vec{v}]dV$$

Expanding the time derivative term and using equation (2.40), one obtains:

$$\iiint \vec{F}\, dV + \iiint [\vec{B} + (\vec{P}.\vec{\nabla})\vec{E}]dV = \int \rho_m \frac{d\vec{v}}{dt}\, dV$$

Expressing the vectors in component form we get:

$$\iiint F_i\, dV + \iiint [B_i + P_l E_{i,l}]dV = \iiint \rho_m \frac{dv_i}{dt} dV \quad i = 1,2,3$$

Using Eq.2.37, the volume integral term corresponding to the near action force can be written in terms of derivatives of stress tensor in the deformed coordinate system. Collecting all the terms, the integral becomes:

$$\iiint \{ B_i + P_l E_{i,l} + \frac{\partial \sigma_{il}}{\partial x_l} \} dV = \iiint \rho_m \frac{dv_i}{dt} dV \qquad i = 1,2,3$$

As this equality is applicable to any arbitrary chosen volume, the conservation of linear momentum equation can be written as:

$$\frac{\partial \sigma_{il}}{\partial x_l} + B_i + P_l E_{i,l} = \rho_m \frac{dv_i}{dt} \tag{2.41}$$

It is obvious that *the peculiar attribute of a polarised dielectric medium is to introduce in the force equilibrium equations a non-linear quantity* $P_l E_{i,l}$ *due to the coupling of polarisation and electric field.* Even in the non-existence of B_i, non-linear $P_l E_{i,l}$ quantity exists. If this non-linear term is discarded then we get the equation of equilibrium for a non-polarised medium. In Vector form above equation is:

$$div\sigma^T + \vec{B} + (\vec{P}.\vec{\nabla})\vec{E} - \rho_m \frac{d\vec{v}}{dt} = 0 \tag{2.42}$$

For uniform electric field $P_l E_{i,l} = (\vec{P}.\vec{\nabla})\vec{E}$ representing coupling of polarisation and electric field vanish.

(c) Conservation of Angular Momentum

The law of conservation of angular momentum can be expressed as:

$$\iiint [R_i \vec{e}_i \times \{P_l E_{j,l} \vec{e}_j + B_j \vec{e}_j\} + (\epsilon_{jkm} \vec{e}_j P_k E_m)] dV + \iiint (\epsilon_{jkm} \vec{e}_j R_k F_m)\, dV$$

$$= D_t \iiint (R_k \vec{e}_k \times \rho_m v_k \vec{e}_k)\, dV$$

Here integration is in the deformed current state and over an arbitrary small volume fragment of the body. It is presumed that per unit volume, intrinsic mass moment of inertia is negligible. The moment arising out of fundamental near action force can be exchanged by a moment out of a fundamental surface force using Eq, 2.37, the above equation becomes:

$$\iiint \left[R_i\vec{e}_i \times \{ P_l E_{j,l}\vec{e}_j + B_j\vec{e}_j\} + (\epsilon_{jkm}\vec{e}_j P_k E_m)\right]dV - \iint \left(\sigma^T\vec{n} \times \vec{R}\right) dA =$$

$$D_t \iiint (\vec{R} \times \rho_m \vec{v})dV \qquad (2.43)$$

Unrevealing the time derivative quantity, RHS of above equation applying transport theorem becomes:

$$D_t \iiint \left(\vec{R} \times \rho_m\vec{v}\right) dV = \iiint \left\{\frac{d}{dt}\rho_m\left(\vec{R} \times \vec{v}\right) - \rho_m\left(\vec{v} \times \vec{R} \right)(\vec{V}.\vec{v})\right\} dV$$

Expanding the time derivative term and rearranging:

$$D_t \iiint \left(\vec{R} \times \rho_m\vec{v}\right) dV$$

$$= \iiint \left\{\left(\vec{R} \times \vec{v}\right)\frac{d\rho_m}{dt} + \rho_m div\,\vec{v} \,) + \rho_m(\vec{v} \right.$$

$$\left. \times \vec{v} \,) + \rho_m(\vec{R} \times \frac{d\vec{v}}{dx}))\right\} dV$$

Applying Eq. 3.40 the time derivative simplifies to:

$$D_t \iiint \left(\vec{R} \times \rho_m\vec{v}\right) dV = \iiint \rho_m \left(\vec{R} \times \frac{d\vec{v}}{dx}\right) dV$$

Substituting in Eq. 3.43, the angular momentum takes the form

$$\iiint [\{ \vec{R} \times \{(\vec{P} .\vec{V})\vec{E} + \vec{B} \} - (\vec{E} \times \vec{P})\}]dV - \iint \left(\sigma^T\vec{n} \times \vec{R}\right) dA = \iiint \rho_m \left(\vec{R} \times \frac{d\vec{v}}{dx}\right) dV \qquad (2.44)$$

The surface integral term can be written as

$$\iint \left(\sigma^T\vec{n} \times \vec{R}\right) dA = -\iiint \{2\,\vec{\omega} + (\vec{R} \times div\sigma^T)\}\, dV$$

Where $\vec{\omega}$ signifies the dual vector of the antiskew part of tensor σ, Substituting the above identity in Eq.2.44, we get

$$\iiint \left[\vec{R} \times \{ div\sigma^T + \vec{B} + (\vec{P} .\vec{V})\vec{E} \,) - \rho_m \frac{d\vec{v}}{dx}\}\right] dV = \iiint (\vec{P} \times \vec{E} + 2\,\vec{\omega})dV$$

$$(2.45)$$

Using Eq.2.42, the first integral vanishes, therefore the second integral must also vanish. Hence the dual vector $\vec{\omega}$ is expressed as:

$$\vec{\omega} = -\frac{1}{2}(\vec{P} \times \vec{E})$$

In component form

$$\omega_1 = -\frac{1}{2}(P_2E_3 - P_3E_2)$$

$$\omega_2 = -\frac{1}{2}(P_3E_1 - P_1E_3)$$

$$\omega_3 = -\frac{1}{2}(P_1E_2 - P_2E_1)$$

Where

$$\vec{\omega} = \omega_1\vec{e}_1 + \omega_2\vec{e}_2 + \omega_3\vec{e}_3$$

The stress tensor can be partitioned into a sum of symmetric tensor and an antisymmetric tensor

$$\sigma = \sigma^S + \sigma^A$$

The antisymmetric tensor σ^A has the components as obtained from dual vector of ant skew vector $\vec{\omega}$

$$\sigma^A = \begin{pmatrix} 0 & -\frac{1}{2}(P_1E_2 - P_2E_1) & \frac{1}{2}(P_3E_1 - P_1E_3 \\ \frac{1}{2}(P_1E_2 - P_2E_1) & 0 & -\frac{1}{2}(P_2E_3 - P_3E_2) \\ -\frac{1}{2}(P_3E_1 - P_1E_3) & \frac{1}{2}(P_2E_3 - P_3E_2) & 0 \end{pmatrix}$$

(2.46)

The distributed moment produced by the coupling of polarisation and electric field manifests the stress tensor as non-symmetric. (Any non-symmetric stress tensor can be expressed as the summation of symmetric and antisymmetric stress tensors.) The skew symmetric portion of the stress tensor as given in Eq. 2.46, is obtainable from the components of the polarisation and the electric field. These elements of the antisymmetric part of the stress tensor are written in indicial notation as:

$$\sigma_{ij}^{A} = \frac{1}{2}(E_i P_j - P_i E_j) \tag{2.47}$$

It is obvious that if polarisation vector aligned with the direction of the electric field, the dual vector $\vec{\omega}$ vanishes and hence the skew symmetric stress tensor also vanishes.

2.5 Constitutive Equations

The principles of thermodynamics i.e. first and second laws will be utilised for obtaining constitutive relations.

2.5.1 First Law of Thermodynamics

The principle of energy conservation of a polarised Continuum consists of contributions from mechanical, electrical and thermal effects. The equation of energy balance for an elemental volume δV in rate form will be:

$$D_t \int (\frac{1}{2} \rho_m v^2 + \rho_m U)dV$$

$$= \int \sigma^T \vec{n}.\vec{v}\, dA + \int \vec{B}.\vec{v}\, dA - \int \vec{Q}.\vec{n}\, dA + \int \rho_m \gamma\, dV$$

$$+ \int \hat{W}_E dV \tag{2.48}$$

Here ρ_m is density, kinetic energy per unit volume is $\frac{1}{2}\rho_m (\vec{v}.\vec{v})$, internal energy per unit volume is $\rho_m U$, work per unit time by the near action forces that are manifested as surface tractions is $\sigma^T\vec{v}.\vec{n}$, the work of gravity force per unit of time is $\vec{v}.\vec{B}$, the rate of heat flux flowing away through the boundary of the infinitesimal volume is $\vec{Q}$, the unit normal pointing out is $\vec{n}$, the rate of heat generated within the volume per unit mass is γ and $\hat{W}_E$ is the rate of work done per unit volume because of the coupling of electric field and polarisation. To establish, the thermodynamic First Law in useable form, various relations are obtained and used in Eq. 2.48.

Using Reynold's transport Theorem and law of mass of conservation relation we have:

$$D_t \iiint \left\{ \frac{1}{2} \rho_m \vec{V}.\vec{V} + \rho_m U \right\} dV$$

$$= \iiint \rho_m \frac{d}{dt} \left(\frac{1}{2} \vec{V}.\vec{V} + U \right) dV$$

$$= \iiint \rho_m \left(\vec{V}.\frac{d\vec{V}}{dt} + \frac{dU}{dt} \right) dV \tag{2.49}$$

Applying Gauss divergence theorem & using vector identities

$$\int \sigma^T \vec{n}.\vec{v}\, dA$$

$$= \int div\, (\sigma \vec{v}) dV$$

$$= \int (\vec{v}.div\sigma^T + \sigma^T.\vec{\nabla}\vec{v})\, dV \tag{2.50}$$

$\sigma^T.\vec{\nabla}\vec{v}$ represent the dot product of tensors, defined for tensors **P** and **Q** as $\boldsymbol{P}.\boldsymbol{Q} =$

$P_{ij}Q_{ij}$

$$\int \vec{Q}.\vec{n}\, dA = \int div\, \vec{Q}\, dV \tag{2.51}$$

In conveying the endowment amount of energy transfer by electric field on the dipole continuum, alteration of electric field with time is ignored. This supposition convey that the effects of the electric field are considered as quasi-static. Power density transferred to the dipole continuum in the deformed current coordinates [3] is:

$$\hat{W}_E = \left[q^+ \left\{ \left(\vec{V} + \frac{d\vec{a}}{dt} \right) . \vec{E}\,(\vec{R} + \vec{a}) \right\} - q^- \{ \vec{E}\,(\vec{R}).(\vec{V})\ \} \right]$$

Here for a dipole continuum, q^+ is the positive charge density and q^- is the negative charge density. It is also eminent that for a dipole continuum charge density, $\rho = q^- = q^+$

Using $\vec{E}(\vec{R}+\vec{a}) = \vec{E}(\vec{R}) + (\vec{\nabla}.\vec{a})\vec{E}(\vec{R})$, we get

$$\hat{W}_E = \left[\rho\left\{\left(\vec{v} + \frac{d\vec{a}}{dt}\right).\vec{E}(\vec{R}) + (\vec{\nabla}.\vec{a})\vec{E}(\vec{R})\right\} - \rho\{\vec{E}(\vec{R}).(\vec{v})\}\right]$$

Omitting terms of higher power of $\vec{a}$ we get $\hat{W}_E = \vec{E}(\vec{R}).\rho\frac{d\bar{a}}{dt} +$

$$\rho(\vec{a}.\vec{\nabla})\vec{E}(\vec{R}).(\vec{v}) \tag{2.52}$$

The first part of the right hand side of above equation is adjusted by using

$$\frac{d}{dt}(\rho\vec{a}) = \rho\frac{d\vec{a}}{dt} + \vec{a}\frac{d\rho}{dt}$$

Since density of polarisation per unit volume, $\rho\vec{a} = \vec{P}$ then obviously

$$\rho\frac{d\bar{a}}{dt} = \frac{d}{dt}(\vec{P}) - \frac{\bar{P}}{\rho}\frac{d\rho}{dt} \tag{2.53}$$

Time derivative of polarisation can be used to manifest $\frac{d\bar{a}}{dt}$. This can be done by using equations of laws of the conservation of charge and mass. Suppose ρ_m is the density of the material; then mass conservation from Eq. 2.40, is

$$\frac{d\rho_m}{dt} + \rho_m\vec{\nabla}.\vec{v} = 0$$

Equation of law of charge conservation can be written as:

$$\frac{d\rho}{dt} + \rho\vec{\nabla}.\vec{v} = 0$$

Comparing equations of charge and mass conservation:

$$\frac{1}{\rho}\frac{d\rho}{dt} = \frac{1}{\rho_m}\frac{d\rho_m}{dt}$$

Substituting in Eq. 3.53, we have

$$\rho\frac{d\bar{a}}{dt} = \frac{d}{dt}(\vec{P}) - \frac{\vec{P}}{\rho_m}\frac{d\rho_m}{dt} \tag{2.54}$$

The polarisation per unit mass is defined as;

$$\vec{\pi} = \frac{\vec{P}}{\rho_m}$$

Substituting in Eq. 2.54, we have:

$$\rho \frac{d\bar{a}}{dt} = \frac{d}{dt}(\rho_m \vec{\pi}) - \vec{\pi}\frac{d\rho_m}{dt}$$

Or

$$\rho \frac{d\vec{a}}{dt} = \rho_m \frac{d\pi}{dt} \tag{2.55}$$

Substituting Eq. 2.55 in Eq. 2.52,

$$\hat{W}_E = \vec{E}(\vec{R}).\rho_m \frac{d\vec{\pi}}{dt} + (\vec{P}.\vec{\nabla})\vec{E}(\vec{R}).(\vec{v}) \tag{2.56}$$

Substituting Eqs, 3.49, 3.50, 3.51and 3.56 in Eq, 3.48, the Eq. 3.48, simplifies to

$$\iiint [\vec{v}.\left\{ \rho_m \frac{d\vec{v}}{dt} - div\sigma^T - \vec{B} - (\vec{\nabla}.\vec{P})\vec{E}(\vec{R})\right\} + \rho_m \frac{dU}{dt} - \sigma^T.\vec{\nabla}\vec{v} + \vec{\nabla}.\vec{Q}$$

$$-\rho_m \gamma - \vec{E}(\vec{R}).\rho_m \frac{d\vec{\pi}}{dt}]\, dV = 0$$

Using Eq. 3.42, the energy balance equation simplifies to

$$\rho_m \frac{dU}{dt} - \sigma^T.\vec{\nabla}\vec{v} + div\,\vec{Q} - \rho_m \gamma - \vec{E}(\vec{R}).\rho_m \frac{d\vec{\pi}}{dt} = 0 \tag{2.57}$$

In component form

$$\rho_m \frac{dU}{dt} = \sigma_{ij}\, v_{j,i} + \rho_m E_i \frac{d\pi_i}{dt} - Q_{i,i} + \rho_m \gamma \tag{2.58}$$

Eq. 3.57 or Eq. 3.58, represent the law of conservation of Energy for a dipole continuum.

Thermodynamics Second law associating the rate of entropy input and entropy generation via heat supply is listed as:

$$\frac{D}{Dt}\iiint \rho_m\, \lambda\, dV \geq \iiint \frac{\rho_m \gamma}{\theta}\, dV - \iint \frac{\vec{Q}}{\theta}.\vec{n}\, dA$$

Where L.H.S depict the entropy generation rate in the system with λ being specific entropy. For absolute temperature θ the R.H.S. quantities emphasize supplied entropy to the system per unit of time.

Using Reynold transport theorem , for L.H.S integral & divergence theorem, for the second integral on R.H.S, the entropy inequality relation can be written as:

$$\iiint \{\frac{d}{dt}(\rho_m \lambda) + \rho_m \lambda\, div\vec{v}\}\, dV \geq \iiint \frac{\rho_m \gamma}{\theta}\, dV - \iiint div\frac{\vec{Q}}{\theta} dV$$

$$\rho_m \iiint \frac{d\lambda}{dt}\, dV + \iiint \lambda\left\{\frac{d\rho_m}{dt} + \rho_m\, div\vec{v}\right\} dV + \iiint div\frac{\vec{Q}}{\theta} dV - \iiint \frac{\rho_m \gamma}{\theta}\, dV \geq 0$$

Using Eq. 2.40, for the second integral, the second law simplifies to

$$\iiint (\, \rho_m \frac{d\lambda}{dt} + div\, \frac{\vec{Q}}{\theta} - \frac{\rho_m \gamma}{\theta})dV \geq 0$$

Or

$$\rho_m \frac{d\lambda}{dt} + div\frac{\vec{Q}}{\theta} - \frac{\rho_m \gamma}{\theta} \geq 0$$

Using $div(\,\varphi\vec{u}) = \varphi div\,\vec{u} + \vec{\nabla}\varphi.\vec{u}$, for the divergence term, with $\varphi = \theta$ and $\vec{u} = \vec{Q}$ the entropy inequality can be written as:

$$\rho_m \frac{d\lambda}{dt} + \frac{1}{\theta}\, div\,\vec{Q} - \frac{1}{\theta^2}\,\vec{\nabla}\,\theta.\vec{Q} - \frac{\rho_m \gamma}{\theta} \geq 0$$

Substituting for $div\,\vec{Q}$ from Eq, 2.58.

$$\rho_m \frac{d\lambda}{dt} + \frac{1}{\theta}(-\rho_m\, \frac{dU}{dt} + \,\sigma_{ij}\, v_{j,i} + \rho\,_m E_i \frac{d\,\pi_i}{dt}) - \frac{1}{\theta^2}\,\, Q_i \theta_{,i} \geq 0 \tag{2.59}$$

The deformed coordinates are used to define quantities involved in Eq. 2.59,

Applying chain rule of differentiation and making use of Eqs. 2.35 and 2.36, we get:

$$v_{j,i} = \frac{\partial v_j}{\partial y_i} = \frac{\partial X_M}{\partial y_i}\frac{\partial v_j}{\partial X_M} = X_{M,i}\frac{\partial}{\partial X_M}\left(\frac{dy_j}{dt}\right) = X_{M,i}\frac{d}{dt}(y_{j,M})$$

$$\theta\rho_m \frac{d\lambda}{dt} - \rho_m\, \frac{dU}{dt} + \,\sigma_{ij}\, X_{M,i}\frac{d}{dt}(y_{j,M}) \,+\, \rho\,_m E_i \frac{d\,\pi_i}{dt} - \frac{1}{\theta}\, Q_i\, \theta_{,i} \geq 0 \tag{2.60}$$

Observing that the temperature θ, the deformation gradient $y_{j,M}$, and the local electric field E_i are measurable variables, Thus defining a thermodynamic function χ known as the

Helmholtz free energy, that is set as: $\chi = U - \lambda\theta - E_i\pi_i$. For the total time derivative taken for χ, we have:

$$\frac{d\chi}{dt} = \frac{dU}{dt} - \lambda\frac{d\theta}{dt} - \theta\frac{d\lambda}{dt} - \pi_i\frac{dE_i}{dt}$$

Substituting for $\frac{dU}{dt}$ in equation (2.60) and simplifying gives the Clausius Duhem inequality:

$$\theta\rho_m\frac{d\lambda}{dt} + \rho_m(-\frac{d\chi}{dt} - \lambda\frac{d\theta}{dt} - \theta\frac{d\lambda}{dt} - \pi_i\frac{dE_i}{dt}) + \sigma_{ij}\, X_{M,i}\frac{d}{dt}(y_{j,M}) + \rho_m E_i\frac{d\,\pi_i}{dt} - \frac{1}{\theta}\,Q_i\,\theta_{,i} \geq 0$$

$$-\rho_m\frac{d\chi}{dt} + \sigma_{ij}\,X_{M,i}\frac{d}{dt}(y_{j,M}) - \rho_m\lambda\frac{d\theta}{dt} - \rho_m\,\pi_i\frac{dE_i}{dt} + \rho_m E_i\frac{d\,\pi_i}{dt} - \frac{1}{\theta}\vec{Q}\,\vec{\nabla}\theta \geq 0$$

(2.61)

Conjecturing Helmholtz free energy function may be conveyed as function of observable variables as:

$$\chi = \chi\,(y_{j,M}, E_i, \theta)$$

Time derivation of this function yields:

$$\frac{d\chi}{dt} = \frac{\partial\chi}{\partial y_{j,M}}\frac{d}{dt}(y_{j,M}) + \frac{\partial\chi}{\partial E_i}\frac{dE_i}{dt} + \frac{\partial\chi}{\partial\theta}\frac{d\theta}{dt} \tag{2.62}$$

Substituting $\frac{d\chi}{dt}$ from Eq.3.62, in Eq.3.61,

$$-\rho_m(\frac{\partial\chi}{\partial y_{j,M}}\frac{d}{dt}(y_{j,M}) + \frac{\partial\chi}{\partial E_i}\frac{dE_i}{dt} + \frac{\partial\chi}{\partial\theta}\frac{d\theta}{dt}) + \sigma_{ij}\,X_{M,i}\frac{d}{dt}(y_{j,M}) - \rho_m\lambda\frac{d\theta}{dt} - \rho_m\,\pi_i\frac{dE_i}{dt} + \rho_m E_i\frac{d\,\pi_i}{dt} - \frac{1}{\theta}\vec{Q}\,\vec{\nabla}\theta \geq 0$$

$$[\sigma_{ij}\,X_{M,i} - \rho_m\frac{\partial\chi}{\partial y_{j,M}}]\frac{d}{dt}(y_{j,M}) - \rho_m\,[\pi_i + \frac{\partial\chi}{\partial E_i}]\frac{dE_i}{dt} - \rho_m[\lambda + \frac{\partial\chi}{\partial\theta}]\frac{d\theta}{dt} - \frac{1}{\theta}\vec{Q}\,\vec{\nabla}\theta \geq 0$$

(2.63)

Some of the bracketed terms in this inequality can be cancelled independently.

Suppose $\frac{d\theta}{dt} = 0$ *and* $\vec{\nabla}\theta = 0$ that is elastic deformation occurs at constant uniform temperature θ. Also for constant electric field and assuming that irrespective of value of $\frac{d}{dt}(y_{j,M})$ above inequality holds, generates constitutive law for stress as following:

$$\sigma_{ij}\, X_{M,i} = \rho_m \frac{\partial \chi}{\partial y_{j,M}} \tag{2.64}$$

As i is dummy index

$$X_{M,i}\, \sigma_{ij} = \rho_m \frac{\partial \chi}{\partial y_{j,M}}$$

Taking inverse of $X_{M,i}$one can write:

$$\sigma_{ij} = \rho_m y_{i,M} \frac{\partial \chi}{\partial y_{j,M}} \tag{2.65}$$

For a similar reasoning, keeping $\vec{\nabla}\theta = 0$ and $\frac{d\theta}{dt} = 0$ yields the constitutive law for polarisation intensity:

$$\pi_i = -\frac{\partial \chi}{\partial E_i} \tag{2.66}$$

Finally, the constitutive relation for entropy is given as:

$$\lambda = -\frac{\partial \chi}{\partial \theta} \tag{2.67}$$

Since χ cannot be any random function *of* θ $y_{j,M}$ and E_i as it must satisfy the invariance principle under rigid body motion for coordinate transformation and thus is scalar invariant function per unit mass of internal energy for deformable polarized continuum. From $\chi = \chi(\theta, y_{j,M}, E_i)$ it is clear that χ is a function of basically 13 variables which consists one of temperature, nine for deformation and three for field. For invariance of coordinate transformation to be valid it must be function of either vector magnitudes, dot product or mixed triple product of vectors, with four vectors under consideration three deformation gradients and one of electric field we can generate fourteen quantities as sum of following, combination four vectors taken one at time C(4,1), combinations of four vectors taken two at a time C(4,2) and combinations of four vectors taken three at a time C(4,3), the first combination is for magnitude of four vectors, the second combination for their dot product and the third combination their mixed triple products yielding 14 quantities, 4 vector magnitudes, 6 dot products and 4 mixed products but according to theorem of invariant functions of several variables, a quantity must be function of only independent variables, out of 14 only 9 are independent and independent ones will be taken as three vector magnitudes and 6 dot product of four vectors, thus asserting the invariance of χ [Tiersten, H.F., "On the non-linear Equations of Electro-thermo-elasticity," International Journal of Engineering Sciences, Vol. 9, No. 7, pp. 587-604, 1971], which are taken as:

$$\acute{C}_{LN} = y_{k,L} y_{k,N}$$

$$W_L = y_{k,L} E_k$$

In undeformed state $\acute{C}_{LN}$, is non zero implying it may be replaced by Green strain tensor given as:

$$\hat{G}_{LN} = \frac{1}{2}(\acute{C}_{LN} - \delta_{LN})$$

Therefore Helmholtz free energy χ may be depicted as a function of quantities described with reference to initial material coordinates as:

$$\chi = \chi(\hat{G}_{LN}, W_L, \theta)$$

Where $\hat{G}_{LN}$ the strain in the initial material coordinates, W_L is the local electric field in the initial material coordinates. Substituting this value of χ in Eqs. 2.65, 2.66 and 2.67, the constitutive equations can be registered as

$$\sigma_{ij} = \rho_m y_{i,L} \frac{\partial \chi}{\partial \hat{E}_{LN}} y_{j,N} + \rho_m y_{i,L} \frac{\partial \chi}{\partial W_L} E_j \qquad (2.68)$$

$$\pi_i = -y_{i,M} \frac{\partial \chi}{\partial W_M} \qquad (2.69)$$

$$\lambda = -\frac{\partial \chi}{\partial \theta} \tag{2.70}$$

Substituting Eq, 3.69, in Eq, 3.68, we get

$$\sigma_{ij} = \rho_m y_{i,L} \frac{\partial \chi}{\partial\, \hat{E}_{LN}} y_{j,N} - \pi_i\, E_j$$

Using the definition of π: $\vec{\pi} = \frac{\vec{P}}{\rho\, m}$

$$\sigma_{ij} = \rho_m y_{i,L} \frac{\partial \chi}{\partial\, \hat{E}_{LN}} y_{j,N} - P_i\, E_j \tag{2.71}$$

The above stress equation depicts that stress tensor comprise of two parts: Symmetric and non-symmetric. Electric field and polarisation interaction gives birth to non-symmetric part.

For infinitesimal deformation gradients the undeformed and deformed positions may be supposed to be nearly identical. This assumption implies any difference between subscripts in upper case and lower case vanish. For ease lower case subscripts will be utilised. $\hat{G}_{LN}$ is exchanged by the conventional Strain tensor ε_{ij} and the scalar W_I by E_i Impetrating these estimates and skipping terms of higher order yield constitutive relations as:

$$\sigma_{ij} = \rho_m \frac{\partial \chi}{\partial\, \varepsilon_{ij}} - P_i\, E_j \tag{2.72}$$

$$\pi_i = -\frac{\partial \chi}{\partial E_i} \tag{2.73}$$

$$\lambda = -\frac{\partial \chi}{\partial \theta} \tag{2.74}$$

Substituting constitutive relations in Eq, 2.63, yields:

$$-\,Q_i\, \theta_{,i} = -\,\frac{1}{\theta} \vec{Q}\, \vec{\nabla} \theta \geq 0$$

It is term corresponding to thermal dissipation. It is a logical fact that heat never flows from lower temperature region to higher temperature region. If the temperature gradient $\theta_{,i}$ is negative then the heat flux rate Q_i has to be positive viz, heat is flowing into the surroundings from the system on the other hand if the temperature gradient is positive then Q_i is negative viz, heat is flowing from the surroundings to the system. As the temperature θ is always positive, the thermal quantity $-\frac{1}{\theta}\vec{Q}\,\vec{\nabla}\theta$ in Eq. 2.63, is always positive.

Here a note about the nature of the term $-Q_i\,\theta_{,i}$ is necessary to be presented. This term is a non-negative quantity, with the value equal to zero when θ is uniform. $\vec{Q}$ and $\vec{\nabla}\theta$ are work conjugates, and thus the term $-\,Q_i\,\theta_{,i}$ will remain unaltered for any rigid body motion. Hence by expressing in initial coordinate system gives:

$$\theta_{,i} = X_{L,i}\theta_{,L}$$

Thus one gets

$$-\,Q_i\,\theta_{,i} = -Q_i\,X_{L,i}\theta_{,L} = -X_{L,i}Q_i\,\theta_{,L} = \bar{Q}_L\,\theta_{,L} \geq 0 \qquad (2.75)$$

$\bar{Q}_L$ is equivalent flux in initial system. For the inequality to be valid $\bar{Q}_L$ has to be an odd function of $\theta_{,L}$. Linear constitutive relation model will thus be $\bar{Q}_L = -K_{LM}\theta_{,M}$ where

tensor K is positive definite and known as thermal conductivity. This relationship (in the deformed coordinate system) leads to:

$$Q_i = y_{i,L}\bar{Q}_L = -\,y_{i,L}K_{LM}\,\theta_{,M} = -\,y_{i,L}\,K_{LM}\,y_{j,M}\,\theta_{,j} = -\,K_{ij}\theta_{,j} \qquad (2.76)$$

The components of tensor K_{ij} are the constituents of thermal conductivity tensor in the current coordinates, Eq. 2.76, is called as heat conduction law as given by Fourier. For isotropic materials there is no favourable direction for heat flow. Enabling, the thermal conductivity tensor to be expressed as $K_{ij} = kI_{ij}$ where k is a non-negative scalar and I_{ij} is unit tensor. Hence in isotropic thermal material Eq, 2.76, can be expressed as:

$$\vec{Q} = -k\vec{\nabla}\theta \qquad (2.77)$$

The above Eq. 2.77, is Fourier law of heat conduction and k is thermal conductivity coefficient.

2.5.2 Governing Equations for Heat Conduction

From first law of thermodynamics Eq. 2.58:

$$\rho_m \frac{dU}{dt} = \sigma_{ij}\, v_{j,i} + \rho_{\,m} E_i \frac{d\,\pi_i}{dt} - Q_{i,i} + \rho_{\,m}\gamma$$

Substituting $U = \chi + \lambda\theta + E_i\pi_i$ in the above equation leads to:

$$\rho_m \frac{d\chi}{dt} = \sigma_{ij}\, v_{j,i} - \rho_m \lambda \frac{d\theta}{dt} - \rho_{\,m}\pi_i \frac{dE_i}{dt} - Q_{i,i} + \rho_{\,m}\gamma \qquad (2.78)$$

Expand $\frac{d\chi}{dt}$ in observable variables and using Eq. 2.62,

$$\frac{d\chi}{dt} = \frac{\partial\chi}{\partial y_{j,M}}\frac{d}{dt}(y_{j,M}) + \frac{\partial\chi}{\partial E_i}\frac{dE_i}{dt} + \frac{\partial\chi}{\partial\theta}\frac{d\theta}{dt}$$

in Eq. 2.78, and using constitutive Eqs. 2.64, 2.65, 2.66 and 2.67 yields

$$Q_{i,i} = \rho_{\,m}\gamma - \rho_{\,m}\theta\frac{d\lambda}{dt} \qquad (2.79)$$

The material derivative of entropy λ can be expressed as:

$$\frac{d\lambda}{dt} = \frac{d}{dt}\left(-\frac{\partial\chi}{\partial\theta}\right) = -\frac{\partial^2\chi}{\partial y_{j,M}\partial\theta}\frac{\partial}{\partial t}(y_{j,M}) - \frac{\partial^2\chi}{\partial\theta\partial E_i}\frac{dE_i}{dt} - \frac{\partial^2\chi}{\partial\theta^2}\frac{\partial\theta}{\partial t}$$

Using Eqs. 2.64, 2.65, 2.66 and 2.67.

$$\frac{d\lambda}{dt} = \frac{-1}{\rho_m}\frac{\partial}{\partial\theta}(\sigma_{ij}\, X_{M,i})\,\frac{\partial}{\partial t}(y_{j,M}) + \frac{d\pi_i}{d\theta}\frac{dE_i}{dt} + \frac{\partial\lambda}{\partial\theta}\frac{\partial\theta}{\partial t} \qquad (2.80)$$

Substituting Eq.2.80 in Eq. 2.79,

$$Q_{i,i} = \rho_{\,m}\gamma + \theta\,\frac{\partial}{\partial\theta}(\sigma_{ij}\, X_{M,i})\,\frac{\partial}{\partial t}(y_{j,M}) - \rho_{\,m}\theta\,\frac{d\pi_i}{d\theta}\frac{dE_i}{dt} - \rho_{\,m}\theta\,\frac{\partial\lambda}{\partial\theta}\frac{\partial\theta}{\partial t}$$

Denoting $C = \frac{\partial\lambda}{\partial\theta}$, the specific heat, along with using heat conduction law from Eq, 2.77 to obtain the complete, coupled heat equation as:

$$-k\Delta\theta = \rho_{\,m}\gamma - \rho_{\,m}\theta\, C\frac{\partial\theta}{\partial t} + \theta\,\frac{\partial}{\partial\theta}(\sigma_{ij}\, X_{M,i})\,\frac{\partial}{\partial t}(y_{j,M}) - \rho_{\,m}\,\theta\,\frac{d\pi_i}{d\theta}\frac{dE_i}{dt}$$

(2.81)

Where Δ denotes Laplacian operator

the couplings involve

(a) Thermomechanical coupling: $\frac{\partial}{\partial\theta}(\sigma_{ij}\, X_{M,i})\,\frac{\partial}{\partial t}(y_{j,M}) - \rho_{\,m}$

(b) Pyroelectric coupling: $\rho_{\,m}\,\theta\,\frac{d\pi_i}{d\theta}\frac{dE_i}{dt}$

In general, coupling quantities both are omitted, to give the equations in standard form that are uncoupled heat conduction relations:

$$\rho_{\,m}\theta\, C\frac{\partial\theta}{\partial t} + -k\Delta\theta = \rho_{\,m}\gamma \qquad (2.82)$$

2.5.3 Constitutive Equations in Useful Form

As quadratic form per unit mass of Helmholtz free energy can be expressed as [Tiersten]:

$$\chi = (2\rho_{\,m})^{-1}(\,C_{mnkl}\varepsilon_{mn}\varepsilon_{kl} - \; b_{mn}E_mE_n + \rho_{\,m}\,C_\theta\theta^2 - 2e_{mnk}E_m\varepsilon_{nk} + 2\alpha_{mn}\varepsilon_{mn}\theta + 2\eta_mE_m\theta) \qquad (2.83)$$

Where different constants are C_{mnkl} elastic, b_{mn} electric susceptibility, C_θ thermal, e_{mnk} piezoelectric, α_{mn} thermoelastic, η_m pyroelectric. Using Eq, 2.83 in equation numbers 2.72, 2.73 and 2.74 the constitutive relations can be obtained. Using Eq, 2.26 and definition of polarisation per unit mass $\vec{\pi} = \frac{\vec{P}}{\rho_{\,m}}$, the constitutive relation for polarisation density in Eq, 2.81 can be expressed in terms of electrical displacement vector $\vec{D}$. The eventual set of constitutive equations are :

$$\sigma_{ij} = C_{ijmn}\varepsilon_{mn} - e_{ijl}E_l - \alpha_{ij}\theta - P_iE_j \qquad (2.84)$$

$$D_i = e_{imn}\varepsilon_{mn} - b_{in}E_n - \eta_i\theta \qquad (2.85)$$

$$\lambda = \alpha_{mn}\varepsilon_{mn} + \; \eta_nE_n + C_\theta\theta \qquad (2.86)$$

$$Q_i = -K_{il}\theta_{,l} \tag{2.87}$$

It is worth noting that for Eq, 2.84, the tensor σ_{ij} can be divided into sum of linear σ_{ij}^{L} and non linear σ_{ij}^{NL} stress tensors, where $\sigma_{ij}^{L} = C_{ijmn}\varepsilon_{mn} - e_{ijl}E_l - \alpha_{ij}\theta$ and $\sigma_{ij}^{NL} = -P_iE_j$

Summary

Starting with Coulombs law the electrostatic equations for a dielectric continuum were derived. Using conservation laws of mass, charge, energy, linear momentum and angular momentum governing equations have been established. Using second law of thermodynamics constitutive equations are established.

The important observations are

The Coupling of electric field and polarization vectors result in non-linear distributed body force in equilibrium equation and distributed body couple (non-linear term) in constitutive equation, which in turn renders the stress tensor non-symmetric. The non-linear term in equilibrium equation vanishes for uniform electric field and if the polarization and electric fields are parallel then the non-linear term in constitutive equation evaporate.

Chapter -III

FEM MODEL OF A SMART BEAM

3.1 Introduction

In the previous chapter the electro-thermo-elastic formulation has been developed for modeling the de-bonding. Formulation involves the development of two sets of equations. They are: equilibrium equations and constitutive relations. The governing equations so developed have been modified by using certain assumptions. The complex equations become simple by assuming the polarisation vector parallel to the electric field vector. This makes the stress tensor symmetric and stress strain relation linear. Now variational formulation relative to the equilibrium equations will be evolved by invoking the virtual work principle in this chapter. Then suitable boundary conditions are obtained and 3-D problem narrowed to certain electro-thermo-elastic bonded and debonded smart beam problems.

3.2 Variational Formulation

Consider a Ferroelectric Continuum in a domain V with boundary S. In order to obtain the variational formulation, the principle of virtual work is invoked and accordingly multiplying Eq. 2.41, with δu_j, Eq. 2.42, with $\delta\,\varphi$ and Eq. 2.82, with $\delta\,\theta$ and integrating over the full piezoelectric domain V gives (neglecting gravity force and assuming absence of heat source under static case):

$$\iiint (\sigma_{il,i} + P_i E_{l,i})\,\delta u_l\,dV + \iiint D_{m,m}\,\delta\phi dV + \frac{1}{\theta_0}\iiint Q_{k,k}\delta\theta\,dV = 0 \tag{3.1}$$

where δu_l is the displacement variation, $\delta\phi$ is the electric potential variation and $\delta\,\theta$ is the temperature variation and θ_0 is the reference temperature.

In Eq. 3.1, steady state heat conduction has been assumed. This decouples the heat conduction and deformation interaction of the body. However deformation and electric potential will be induced due to temperature difference. In this study temperature difference θ is not an unknown variable but is an external load, Eq. 3.1, can be written as:

$$\iiint \frac{\partial(\sigma_{il}\delta u_i)}{\partial x_l} dV - \iiint \sigma_{il} \frac{\partial \delta u_i}{\partial x_l} dV + \iiint \frac{\partial(D_m \delta\phi)}{\partial x_m} dV - \iiint D_m \frac{\partial \delta\phi}{\partial x_m}$$
$$+ \frac{1}{\theta_0}\iiint \frac{\partial(Q_k \delta\theta)}{\partial x_k} dV - \frac{1}{\theta_0}\iiint Q_k \frac{\partial \delta\theta}{\partial x_k} dV + \iiint P_l E_{i,l}\, \delta u_i dV = 0$$

(3.2)

The term $\iiint P_l E_{i,l}\, \delta u_i dV$ constitute virtual work because of dispense non-linear force generated by the coupling of the vector electric field and the polarisation vector. Also, stress tensor σ_{il} is non-symmetric and is given as: $\sigma_{il} = \sigma_{ij}^{L} + \sigma_{ij}^{NL}$

The linear σ_{ij}^{L} and non linear σ_{ij}^{NL} stress tensors from Eq. 3.84 are given as:

$\sigma_{ij}^{L} = C_{ijmn}\varepsilon_{mn} - e_{ijl}E_l - \alpha_{ij}\theta$ and $\sigma_{ij}^{NL} = -P_i E_j$

Separating the linear and nonlinear parts in Eq. 3.2, we get:

$$\iiint \frac{\partial(\sigma_{il}\delta u_i)}{\partial x_l} dV - \iiint \sigma_{il}^{L} \frac{\partial \delta u_i}{\partial x_l} dV + \iiint \frac{\partial(D_m \delta\phi)}{\partial x_m} dV - \iiint D_m \frac{\partial \delta\phi}{\partial x_m} + \frac{1}{\theta_0}\iiint \frac{\partial(Q_k \delta\theta)}{\partial x_k} dV -$$
$$\frac{1}{\theta_0}\iiint Q_k \frac{\partial \delta\theta}{\partial x_k} dV - \iiint P_i E_j \frac{\partial \delta u_i}{\partial x_j} dV + \iiint P_i E_{i,l}\delta u_i dV = 0 \qquad (3.3)$$

Using Gauss divergence theorem and $\frac{\partial \delta\phi}{\partial x_m} = -\delta E_m$ the equation simplifies to

$$\iiint \sigma_{il}^{L}\,\delta\varepsilon_{il}dV - \iiint D_m\,\delta E_m dV + \frac{1}{\theta_0}\iiint Q_k \frac{\partial\delta\theta}{\partial x_k}dV + \iiint P_i\,E_{i,l}\delta u_i dV -$$

$$\iiint P_i\,E_l\delta\frac{\partial u_i}{\partial x_l}dV = \iint(\sigma_{il}\,n_l\delta u_i + D_i n_i\delta\phi + \frac{1}{\theta_0}Q_i n_i\delta\theta)dA \tag{3.4}$$

From the right hand side it is obvious $\sigma_{il}n_l = T_i$ the boundary traction applied, $D_l n_l = q$ surface charge density and $Q_l n_l = Q_n$ vector heat flux perpendicular to boundary.

$$\iiint \sigma_{il}^{L}\,\delta\varepsilon_{il}dV - \iiint D_m\,\delta E_m dV + \frac{1}{\theta_0}\iiint Q_k \frac{\partial\delta\theta}{\partial x_k}dV + \iiint P_i\,E_{i,l}\delta u_i dV - \iiint P_i\,E_l\delta\frac{\partial u_i}{\partial x_l}dV = \iint T_i\,\delta u_i dA + \iint q\,\delta\phi\,dA + \frac{1}{\theta_0}\iint Q_n\,\delta\theta dA \tag{3.5}$$

The following observations are made from the variational formulation:

(i) The terms $\iiint P_i\,E_{i,l}\delta u_i dV$ *and* $\iiint P_i\,E_l\delta\frac{\partial u_i}{\partial x_l}dV$ constitute virtual work associated with body force (non-linear) and stress (non-linear) associated with interaction of vector electric field with the polarization vector.

(ii) For the right hand, the necessary boundary conditions are grabbed by inflicting identifications on the primary or secondary variables.

Primary	u_i	ϕ	θ
Secondary	T_i	q	Q_n

On the domain, either the electric potential ϕ is given or the electric charge q is zero. Moreover if polarisation vector is parallel to electric field vector then $\vec{P}\times\vec{E}$ is zero. Implementing these conditions one can simplify Eq. 3.5, to obtain problem of electro-elastic origin with the additional assumption of considering the body force (non-linear) of polarization vector to be infinitesimally small

$$\iiint \sigma_{ij}^{L}\, \delta\varepsilon_{ij} dV - \iiint D_m\, \delta E_m dV = \iint T_i\, \delta u_i dA \quad (3.6)$$

3.3 Smart cantilever beam Model

For the cantilever beam shown in Fig 3.1, with global system of coordinate with x-axis in the axial (length) direction, y-axis in depth direction and z-axis along thickness the cantilever beam is assumed to consist of an aluminium metallic central (host) with ferroelectric patches PZT-5H brazed on the bottom and top surfaces. Aluminium is isotropic and PZT-5H is transversely isotropic material. Considering the polarisation axis to be the thickness z direction, the constitutive equation for PZT-5H for linear electro-elastic behaviour is

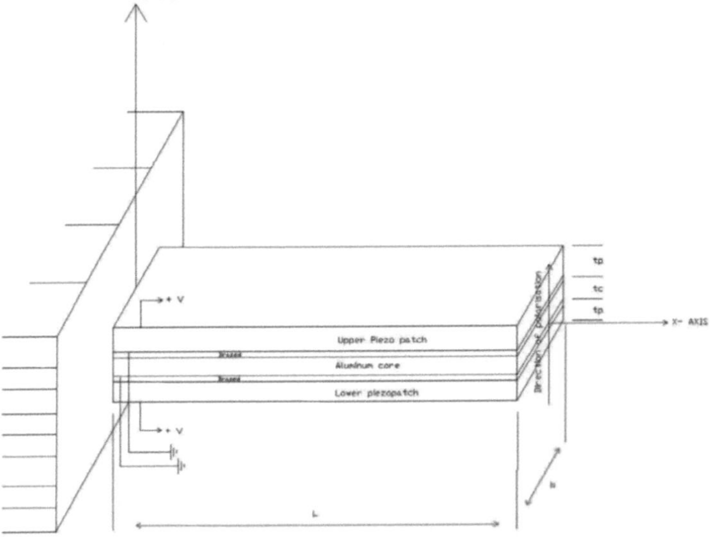

Figure 3.1: Smart Cantilever Beam for Extensional Sensing & Actuation

$$\{ D_x\ D_y\ D_z\ \sigma_x\ \sigma_y\ \sigma_z\ \sigma_{yz}\ \sigma_{xz}\ \sigma_{xy}\}^T$$

$$= \begin{Bmatrix} \epsilon_1^S & 0 & 0 & 0 & 0 & 0 & 0 & e_{15} & 0 \\ 0 & \epsilon_1^S & 0 & 0 & 0 & 0 & e_{15} & 0 & 0 \\ 0 & 0 & \epsilon_3^S & e_{31} & e_{31} & e_{33} & 0 & 0 & 0 \\ 0 & 0 & -e_{31} & c_{11}^E & c_{12}^E & c_{13}^E & 0 & 0 & 0 \\ 0 & 0 & -e_{31} & c_{12}^E & c_{11}^E & c_{13}^E & 0 & 0 & 0 \\ 0 & 0 & -e_{33} & c_{13}^E & c_{13}^E & c_{33}^E & 0 & 0 & 0 \\ 0 & -e_{15} & 0 & 0 & 0 & 0 & c_{44}^E & 0 & 0 \\ -e_{15} & 0 & 0 & 0 & 0 & 0 & 0 & c_{44}^E & 0 \\ 0 & 0 & 0 & 0 & 0 & 0 & 0 & 0 & c_{66}^E \end{Bmatrix} \begin{Bmatrix} E_x \\ E_y \\ E_z \\ \epsilon_x \\ \epsilon_y \\ \epsilon_z \\ \gamma_{yz} \\ \gamma_{zx} \\ \gamma_{xy} \end{Bmatrix}$$

(3.7)

Constants $\epsilon_j^S, e_{lm}\ and\ c_{mn}^E$ refer to components of permittivity, piezoelectric and elastic tensors respectively superscript s and E implying constant strain and electric field conditions. The values of ϵ_1^S and ϵ_3^S for PZT-5H are $1.5045 \times 10^{-8}\, C^2/Nm^2$ and $1.30095 \times \frac{10^{-8} C^2}{Nm^2}$ $D_i\ and\ E_i$ refer to constituents of electric displacement vector and electric field vector respectably and finally $\sigma_{ij}, and\ \epsilon_l\ with\ \gamma_{lm} = 2\epsilon_{lm}$ refer to engineering stress and engineering strain tensor components.

The assumptions for the investigation of the smart cantilever beam are:

(i) Aluminium metallic central (host) and ferroelectric patches PZT-5H brazed on the bottom and top surfaces are in plane strain conditions for x z-plane

(ii) As the beam is not subjected to loading along z-direction, normal stress in z-direction is zero

(iii) The material surfaces at the interface are initially ideally bonded and finally debonding occurs in middle one-third of the span.

These assumptions will be used to further reduce the constitutive relationships given by Eq.3.7, for each specific case study of smart beam.

3.3.1 Displacement and Electric Potential Fields for Layer-by-layer Beam Model

The displacement variables taken for smart cantilever beam investigation are: longitudinal displacement u, displacement along breadth v, displacement in thickness direction w for aluminium metallic central (host) as well as ferroelectric patches and variable electric potential ϕ for ferroelectric patches. Plane strain condition will exist for smart cantilever beam if displacement components and the electric potential are functions of x and z. Utilising higher order bending theory of beams, the axial displacement component and potential are to be expanded in terms of powers of z. The axial displacement components in each layer are expanded upto cubic powers of z, while the electric potential is expanded as a quadratic expression in terms of z.

The displacement field for all layers are supposed to be

$$u^l(x,z) = \sum_{i=0}^{3} z^i\, u_i^l \tag{3.8}$$

$$v^l(z,x) = 0 \tag{3.9}$$

$$w^l(z,x) = w_0[x] \tag{3.10}$$

$u^l(x,z)$ is global displacement variation in x direction, $v^l(x,z)$ is same along y and $w^l(x,z)$ for the z direction for the lth layer.

The electric potential of the ferroelectric patches PZT-5H of the beam in each layer is given by:

$$\phi^l(x,z) = \sum_{i=0}^{2} z^i\, \phi_i^l(x) \tag{3.11}$$

3.3.2 Global and Intrinsic Coordinates

Z coordinate in the global system will be converted to intrinsic $\overline{z}$ coordinate utilizing the substitution $z_c^l = z - \overline{z}$ to simplify the implementation of the model given above.

Therefore in terms of local (intrinsic) coordinates, Eq. 3.11, becomes:

$$\phi^l(\overline{z}, x) = \{\phi_0^l(x) + z_c^l\phi_1^l(x) + (z_c^l)^2\phi_2^l(x)\} + \overline{z}\,\{\phi_1^l(x) + 2z_c^l\phi_2^l(x)\} + (\overline{z})^2\phi_2^l(x) \tag{3.12}$$

Here $\overline{z}$ for particular layer is the intrinsic coordinate and $z_c^l = \frac{z_b^l + z_t^l}{2}$ the span between the centre of the beam up to the centre of a lth layer. z_b^l and z_t^l are the bottom and top z-coordinates of the lth layer, respectively. Eq, 3.12, can be recast in the local coordinate $\overline{z}$ as:

$$\overline{\phi}^l(\overline{z}, x) = \sum_{i=0}^{2}(\overline{z})^i\overline{\phi}_i^l(x) = \overline{\phi}_0^l(x) + \overline{z}\,\overline{\phi}_1^l(x) + (\overline{z})^2\overline{\phi}_2^l(x) \tag{3.13}$$

Proceeding on same approach, longitudinal displacement u in terms of $\overline{z}$ is:

$$\overline{u}^l(x, \overline{z}) = \sum_{i=0}^{3}(\overline{z})^i\,\overline{u}_i^l(x) \tag{3.14}$$

Layer-wise potential and displacement is expressed in local coordinates. The inter layer continuity of the potential and displacement is the primary consideration in this type of modeling.

3.4 Constitutive Equations for Smart Beam

A beam can be made smart in two ways, either by brazing (or joining by adhesive layers) the piezo layers on the upper and lower side of the metallic host material and the other way is that piezo core can be sandwiched between two metallic layers (sandwiched beam). In the sandwiched beam the piezo material core has the polarisation axis parallel to the axial direction of the beam, sandwiched beam leads to shear actuation of the cantilever beam when an electric potential (in z-direction) is applied across the piezo core. In surface mounted smart beam the polarisation axis of both piezo patches points in the transverse positive z direction and thus corresponds to extensional actuation for applied electric potential in z-direction across the piezo-patches.

For several cases the simplified constitutive equations, are separately obtained for the piezo and metallic layers. The various cases are as follows:

1. For linear electro-elastic analysis in extensional mode actuation with ferroelectric outer layers and aluminium core-layer in surface mounted smart beam.
2. For linear electro-thermo-elastic analysis in extensional mode actuation with ferroelectric outer layers and aluminium core layer in surface mounted smart beam.
3. For linear electro-elastic analysis in shear mode actuation with ferroelectric core layer and aluminium outer layers in sandwiched smart beam.
4. For non-linear electro-elastic analysis in extensional mode actuation with ferroelectric outer layers and aluminium centre host layer in surface mounted smart beam.

3.4.1 Linear electro-mechanical problem for surface mounted smart cantilever beam

The simplified material matrix is obtainable from the basic relationship given by Eq. 3.7. Applying the linear electro-mechanical formulation for the ferroelectric outer layers for smart beam in extensional mode actuation. Let us assume layers of beam to be in a state of plane strain with transverse stress σ_z equal to zero. Further, the polarisation axis of the ferroelectric layer is in thickness direction pointing outwards. Enforcing the conditions $\gamma_{yz'} = \gamma_{xy} = \epsilon_y = 0$ and $\sigma_z = 0$ for every layer the resultant simplified constitutive equations are:

$$\sigma_x = c^*\epsilon_x - e^*E_Z \tag{3.15}$$

$$\sigma_{xz} = -e_{15}E_x + c_{44}\gamma_{xz} \tag{3.16}$$

$$D_x = \epsilon_1^S E_x + e_{15}\gamma_{xz} \tag{3.17}$$

$$D_y = \epsilon_1^S E_y + e_{15}\gamma_{yz} \tag{3.18}$$

$$D_z = \epsilon_3^* E_z + e^*\epsilon_x \tag{3.19}$$

Where $c^* = (c_{11} - \frac{c_{13}^2}{c_{33}})$ for both metallic and piezo materials,

$e^* = (e_{31} - \frac{c_{13}}{c_{33}}e_{33})$ and $\epsilon_3^* = (\epsilon_3^S + \frac{e_{33}^2}{c_{33}})$ for piezo materials.

Linear Smart Beam Problem

Applying a tip load F at the free end of the smart cantilever beam, variational formulation using Eq. 4.6, unfolds as:

$$\iiint(\sigma_x\,\delta\epsilon_x + \sigma_{xz}\delta\gamma_{yz})dV - \iiint(D_x\delta E_x + D_z\delta E_z)dV = F\delta w]_{x=L} \tag{3.20}$$

Using the abridged constitutive equations from Eqs. 2.15 - 3.19 and swapping E_m $with -\frac{\partial\phi}{\partial x_m}$ Eq. 3.20, unfolds into:

$$\iiint \{(c^*\epsilon_x + e^*\frac{\partial \phi}{\partial z}) \delta\epsilon_x + (e_{15}\frac{\partial \phi}{\partial x} + c_{44}\gamma_{xz})\delta\gamma_{yz}\}dV - \iiint [\{ (\epsilon_1^s(-\frac{\partial \phi}{\partial x}) + e_{15}\gamma_{xz}\}\delta(-\frac{\partial \phi}{\partial x}) + \{\epsilon_3^*(-\frac{\partial \phi}{\partial z}) + e^*\epsilon_x\}\delta(-\frac{\partial \phi}{\partial z})]dV = F\delta w]_{x=L} \tag{3.21}$$

3.4.2 Problem: Constitutive Relations for linear electro-thermo-mechanical investigation of a surface mounted smart beam

The thermal effect is to be included in the smart beam model in addition to electro-elastic phenomenon for this case. For simplicity, assuming known uniform temperature distribution on the beam a priori from the solution of heat conduction equation. Thus, the temperature difference (between the beam and ambient temperatures) Δ T is given for all material points on the beam. Under thermal excitation, the reduced constitutive equations will be reformulated, starting from the generalized electro-thermo-elastic constitutive relations for PZT-5H (modified version of Eq. 3.7,) which are as follows:

$$\begin{Bmatrix} D_x \\ D_y \\ D_z \\ \sigma_x \\ \sigma_y \\ \sigma_z \\ \sigma_{yz} \\ \sigma_{xz} \\ \sigma_{xy} \end{Bmatrix} = \begin{Bmatrix} \epsilon_1^s & 0 & 0 & 0 & 0 & 0 & 0 & e_{15} & 0 \\ 0 & \epsilon_1^s & 0 & 0 & 0 & 0 & e_{15} & 0 & 0 \\ 0 & 0 & \epsilon_3^s & e_{31} & e_{31} & e_{33} & 0 & 0 & 0 \\ 0 & 0 & -e_{31} & c_{11}^E & c_{12}^E & c_{13}^E & 0 & 0 & 0 \\ 0 & 0 & -e_{31} & c_{12}^E & c_{11}^E & c_{13}^E & 0 & 0 & 0 \\ 0 & 0 & -e_{33} & c_{13}^E & c_{13}^E & c_{33}^E & 0 & 0 & 0 \\ 0 & -e_{15} & 0 & 0 & 0 & 0 & c_{44}^E & 0 & 0 \\ -e_{15} & 0 & 0 & 0 & 0 & 0 & 0 & c_{44}^E & 0 \\ 0 & 0 & 0 & 0 & 0 & 0 & 0 & 0 & c_{66}^E \end{Bmatrix} \begin{Bmatrix} E_x \\ E_y \\ E_z \\ \epsilon_x \\ \epsilon_y \\ \epsilon_z \\ \gamma_{yz} \\ \gamma_{zx} \\ \gamma_{xy} \end{Bmatrix} - \begin{Bmatrix} 0 \\ 0 \\ -\Lambda_3 \\ \beta_1 \\ \beta_1 \\ \beta_3 \\ 0 \\ 0 \\ 0 \end{Bmatrix} \Delta T \tag{3.22}$$

Since each layer in the beam is under plane strain conditions and the transverse stress vanishes; i. e, $\gamma_{yz}, = \gamma_{xy} = \epsilon_y = 0$ and $\sigma_z = 0$ in each layer, the layer wise simplified Constitutive equations are:

$$\sigma_x = c^*\epsilon_x - e^*E_z + \beta^\alpha \Delta T \tag{3.23}$$

$$\sigma_{xz} = -e_{15}E_x + c_{44}\gamma_{xz} \tag{3.24}$$

$$D_x = \epsilon_1^s E_x + e_{15}\gamma_{xz} \tag{3.25}$$

$$D_y = \epsilon_1^s E_y + e_{15}\gamma_{yz} \tag{3.26}$$

$$D_z = \epsilon_3^* E_z + e^*\epsilon_x + \beta^D \Delta T \tag{3.27}$$

Where $\beta^\alpha = \left(\frac{c_{13}}{c_{33}}\beta_3 - \beta_1\right)$, $c^* = \left(c_{11} - \frac{c_{13}^2}{c_{33}}\right)$ *and* $\beta^D = \left(\frac{e_{33}}{c_{33}}\beta_3 + \Lambda_3\right)$ for both metallic and piezo materials; $e^* = (e_{31} - \frac{c_{13}}{c_{33}}e_{33})$ and $\epsilon_3^* = (\epsilon_3^s + \frac{e_{33}^2}{c_{33}})$ for piezo materials.

The variational formulation for the smart cantilever beam under tip load F, from Eq. 3.20, using these simplified constitutive equations for an electro-thermo-mechanical cantilever beam are:

$$\iiint\{(c^*\epsilon_x + e^*\frac{\partial\phi}{\partial z} + \beta^\alpha\Delta T)\,\delta\epsilon_x + (e_{15}\frac{\partial\phi}{\partial x} + c_{44}\gamma_{xz})\delta\gamma_{yz}\}dV - \iiint[\{ (\epsilon_1^s(-\frac{\partial\phi}{\partial x}) + e_{15}\gamma_{xz}\}\delta(-\frac{\partial\phi}{\partial x}) + \{\epsilon_3^*(-\frac{\partial\phi}{\partial z}) + e^*\epsilon_x + \beta^D\Delta T\}\delta(-\frac{\partial\phi}{\partial z})\,]dV = F\delta w]_{x=L} \tag{3.28}$$

By transporting the known temperature gradient terms to the right hand side Eq. 3.28, is refashioned as:

$$\iiint\{(c^*\epsilon_x + e^*\frac{\partial\phi}{\partial z})\,\delta\epsilon_x + (e_{15}\frac{\partial\phi}{\partial x} + c_{44}\gamma_{xz})\delta\gamma_{yz}\}dV - \iiint[\{ (\epsilon_1^s(-\frac{\partial\phi}{\partial x}) + e_{15}\gamma_{xz}\}\delta(-\frac{\partial\phi}{\partial x}) + \{\epsilon_3^*(-\frac{\partial\phi}{\partial z}) + e^*\epsilon_x\}\delta(-\frac{\partial\phi}{\partial z})\,]dV = F\delta w]_{x=L} - \iiint \beta^\alpha\Delta T\delta\epsilon_x\,dV + \iiint \beta^D\Delta T\delta\left(-\frac{\partial\phi}{\partial z}\right)dV \tag{3.29}$$

3.4.3 Problem: Constitutive Relation for Linear electro-elastic analysis of a sandwiched smart beam

The sandwiched beam Fig. 4.2, for electro-elastic analysis considered in this case, the piezo layer forms the core and the metallic layers are arranged and located at the upper and the lower sides of metallic host. The smart cantilever beam is actuated by a shearing action of the piezo core and hence this actuation mechanism is known as shear actuation.

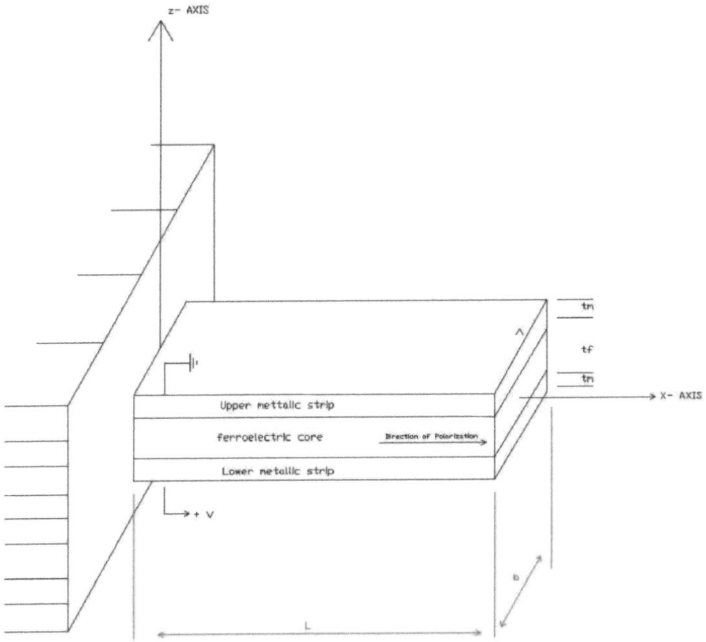

Figure 3.2: Smart Cantilever Beam for Shear Actuation

To achieve shear actuation, the polarisation axis of the core is taken along the longitudinal x- axis. The constitutive relationship for the core, can be obtained from Eq. 3.7, by switching the x and z indices, i.e. indices 1 and 3 are switched. Therefore in Eq.3.7, x axis will be referred by index 3 and z axis will be referred by the index 1. As in the other cases, each layer is assumed to be in plane strain state and the transverse stress is equal to zero. This leads to the following changes in the constitutive equations:

Transverse stress: $\sigma_z = \sigma_1 = 0$

Which implies that $-e_{31}E_3 + c_{11}\epsilon_1 + c_{12}\epsilon_2 + c_{13}\epsilon_3 = 0$ (3.30)

Where $\epsilon_1 = \epsilon_z, \epsilon_2 = \epsilon_y$ and $\epsilon_3 = \epsilon_x$. Further, the assumption of plane strain for x-z plane implies that $\gamma_{yz}, = \gamma_{xy} = \epsilon_y = 0$ using these conditions in Eq.3.30, yields

$$\epsilon_z = \frac{e_{31}}{c_{13}}E_x - \frac{c_{13}}{c_{11}}\epsilon_x \quad (3.31)$$

Noting that $\epsilon_y = 0$ and using the expression for ϵ_z from Eq. 3.31, gives the axial stress σ_x as

$$\sigma_x = -e_{33}E_x + c_{13}\left(\frac{e_{31}}{c_{13}}E_x - \frac{c_{13}}{c_{11}}\epsilon_x\right) + c_{33}\epsilon_x \quad (3.32)$$

Which can be rewritten as $\sigma_x = c^*\epsilon_x - e^*E_x$ (3.33)

Where $c^* = (c_{33} - \frac{c_{13}^2}{c_{11}})$ for both metallic and piezo materials, $e^* = (e_{33} - \frac{c_{13}}{c_{11}}e_{31})$ for the piezo layer. Similarly the other simplified constitutive relations will be obtained as

$$\sigma_{xz} = -e_{15}E_z + c_{44}\gamma_{xz} \quad (3.34)$$

$$D_x = D_3 = \epsilon_3^s E_x + e_{31}\epsilon_x + e_{33}\epsilon_x$$

$$D_y = 0$$

$$D_z = e_1^s E_z + e_{15}\gamma_{xz}$$

Substituting the expression $\epsilon_z = \frac{e_{31}}{c_{13}}E_x - \frac{c_{13}}{c_{11}}\epsilon_x$ from Eq. 3.31, in D_x

$$D_x = \epsilon_3^S E_x + e_{31}\left(\frac{e_{31}}{c_{13}}E_x - \frac{c_{13}}{c_{11}}\epsilon_x\right) + e_{33}\epsilon_x$$

Which can be rewritten as $D_x = \epsilon_1^* E_x + e^*\epsilon_x$ (3.35)

Where $\epsilon_1^* = \left(\epsilon_3^s + \frac{c_{31}^2}{c_{11}}\right)$ $and\ e^* = (e_{33} - \frac{c_{13}}{c_{11}} e_{31})$ for the piezo layer. Note that the expressions for σ_{xz} and D_z remain unchanged.

A tip load of magnitude F is supposed to act in negative Z direction on our smart cantilever beam, and substituting the reduced constitutive relation in Eq.4.20, gives:

$$\iiint \{(c^* \epsilon_x + e^* \frac{\partial\phi}{\partial z}) \delta\epsilon_x + (e_{15} \frac{\partial\phi}{\partial x} + c_{44}\gamma_{xz})\delta\gamma_{yz}\}dV - \iiint [\{ (\epsilon_1^*(-\frac{\partial\phi}{\partial x}) + e^*\epsilon_x\}\delta(-\frac{\partial\phi}{\partial x}) + \{\epsilon_1^s(-\frac{\partial\phi}{\partial z}) + e_{15}\gamma_{xz}\}\delta(-\frac{\partial\phi}{\partial z})]dV = F\delta w]_{x=L} \tag{3.36}$$

3.4.4 Sensing and Actuation

Simplified linear constitutive relations were derived for the piezo and the metallic layers in smart beams. These relations when implemented in Eq. 3.20, unfolds into the desired variational formulation for a smart beam under a tip load. For a mechanical loading, one could use the piezo patches to comprehend the mechanical deformations by measuring the induced electric potential distribution. The magnitude of the induced electric potential can be ascertained, by employing electrodes on the upper and lower surfaces of the piezo layers, which in turn is an indication of the mechanical deformation.

However, the piezo layers of the smart beam will be subjected to an externally applied potential difference, for the case of actuation, which will cause mechanical straining of the piezo patches. This effect will cause bending of the beam. Since no external mechanical load is applied F=0, Eq. 3.20, will have a zero term on right hand side (i.e. no external mechanical load term). Hence the problem will be driven by the known amounts of the potential ϕ on the upper and lower surfaces of the piezo patches.

Adding the nonlinear forces and moments because of the interaction of polarisation vector and electric field vector, all the cases given above can be re-examined. In almost all the studies reported in the literature on the application of smart structures, these nonlinear effects have not been identified. In the following section a detailed variational formulation for the smart beam, including the effects of the nonlinearity, is presented.

3.4.5 Nonlinear electro-elastic analysis of a surface mounted beam

The interaction of polarisation and electric field introduces two types of non-linearities in the smart structure, namely (i) a distributed non-linear force in equilibrium Eq. 2.41, this non-linearity arises due to variable electric field in piezo material and (ii) non-linear constitutive relation between stress-strain-electric field-polarisation Eq. 2.84, this non-linearity is due to the presence of distributed body couple, Each case is presented separately in the following.

(a) Non-linearity due to Distributed Force

In this study, it is assumed that $E_y = 0, \frac{\partial E_x}{\partial y} = 0, \frac{\partial E_y}{\partial y} = 0, \frac{\partial E_z}{\partial y} = 0, \frac{\partial E_y}{\partial z} = 0, \frac{\partial E_y}{\partial x} = 0$.

Thus, in the variational formulation given by Eq. 3.20, the effect of the distributed body force because of polarisation vector and electric field vector interaction, given by $\int (P_i E_{i,j}) \delta u_i dV$, is to be added. This contribution can be written in an expanded form as,

$$\iota^{NL-FORCE} = \iiint P_x \frac{\partial E_x}{\partial x} \delta u_x dV + \iiint P_y \frac{\partial E_x}{\partial y} \delta u_x dV + \iiint P_z \frac{\partial E_x}{\partial z} \delta u_x dV$$
$$+ \iiint P_x \frac{\partial E_y}{\partial x} \delta u_y dV + \iiint P_y \frac{\partial E_y}{\partial y} \delta u_y dV + \iiint P_z \frac{\partial E_y}{\partial z} \delta u_y dV$$
$$+ \iiint P_x \frac{\partial E_z}{\partial x} \delta u_z dV + \iiint P_y \frac{\partial E_z}{\partial y} \delta u_z dV + \iiint P_z \frac{\partial E_z}{\partial z} \delta u_z dV \tag{3.37}$$

Revising the above contribution in terms of the electric displacement using

$$\vec{D} = \varepsilon_0 \vec{E} + \vec{P}$$

$$\iota^{NL-FORCE} = \iiint (D_x - \epsilon_0 E_x)\frac{\partial E_x}{\partial x}\delta u_x dV + \iiint P_y\,(0)\delta u_x dV$$

$$+ \iiint (D_z - \epsilon_0 E_z)\frac{\partial E_x}{\partial z}\delta u_x dV + \iiint P_x(0)\,\delta u_y dV + \iiint P_y(0)\,\delta u_y dV$$

$$+ \iiint P_z(0)\,\delta u_y dV + \iiint (D_x - \epsilon_0 E_x)\frac{\partial E_z}{\partial x}\delta u_z dV + \iiint P_y(0\,)\,\delta u_z dV$$

$$+ \iiint (D_z - \epsilon_0 E_z)\frac{\partial E_z}{\partial z}\delta u_z dV$$

$$\iota^{NL-FORCE} = \iiint \{(D_x - \epsilon_0 E_x)\frac{\partial E_x}{\partial x} + (D_z - \epsilon_0 E_z)\frac{\partial E_x}{\partial z}\}\delta u_x dV + \iiint \{(D_x - \epsilon_0 E_x)\frac{\partial E_z}{\partial x}$$

$$+ (D_z - \epsilon_0 E_z)\frac{\partial E_z}{\partial z}\}\delta u_z dV$$

(3.38)

Using the simplified constitutive relations for the electro-elastic beam, given by Eqs. 3.15-3.19, in the above expression yields:

$$\iota^{NL-FORCE} = \iiint [\{(\epsilon_0 - \epsilon_1^S)\frac{\partial \phi}{\partial x} + e_{15}\gamma_{xz}\}\frac{\partial^2 \phi}{\partial x^2}$$

$$+ \left\{(\epsilon_0 - \epsilon_1^*)\frac{\partial \phi}{\partial z} + e^*\epsilon_x\right\}\frac{\partial^2 \phi}{\partial x \partial z}]\,\delta u_x dV + \iiint [\{(\epsilon_0 - \epsilon_1^*)\frac{\partial \phi}{\partial z} + e^*\epsilon_x\right\}\frac{\partial^2 \phi}{\partial z^2}$$

$$+ \{(\epsilon_0 - \epsilon_1^S)\frac{\partial \phi}{\partial x} + e_{15}\gamma_{xz}\}\frac{\partial^2 \phi}{\partial x \partial z}\delta u_z dV$$

(3.39)

(b) Contribution due to nonlinear part of stress

From Eqs. 4.4 and 4.5, it is clear that for the smart beam the nonlinear term due to stress components is $\iiint P_i\, E_l \delta \frac{\partial u_i}{\partial x_l} dV$

$$\iota^{NL-\sigma} = \iiint P_x E_x \delta \frac{\partial u_x}{\partial x} dV + \iiint P_x E_y \delta \frac{\partial u_x}{\partial y} dV \iiint P_x E_z \delta \frac{\partial u_x}{\partial z} dV$$
$$+ \iiint P_y E_x \delta \frac{\partial u_y}{\partial x} dV + \iiint P_y E_y \delta \frac{\partial u_y}{\partial y} dV + \iiint P_y E_z \delta \frac{\partial u_y}{\partial z} dV$$
$$+ \iiint P_z E_x \delta \frac{\partial u_z}{\partial x} dV + \iiint P_z E_y \delta \frac{\partial u_z}{\partial y} dV + \iiint P_z E_z \delta \frac{\partial u_z}{\partial z} dV$$

(3.40)

Using the assumptions and the simplified constitutive relations as given in Eqs. 3.14 - 3.19, can be written as:

$$\iota^{NL-\sigma} = \iiint [\{(\epsilon_1^s - \epsilon_0)\frac{\partial \phi}{\partial x} + e_{15}\gamma_{xz}\}\left(\frac{\partial \phi}{\partial x}\delta\frac{\partial u_x}{\partial x} + \frac{\partial \phi}{\partial z}\delta\frac{\partial u_x}{\partial z}\right) dV + \iiint \{(\epsilon_1^* - \epsilon_0)\frac{\partial \phi}{\partial z}$$
$$+ e^*\epsilon_x\}(\frac{\partial \phi}{\partial x}\delta\frac{\partial u_x}{\partial x} + \frac{\partial \phi}{\partial z}\delta\frac{\partial u_x}{\partial z})dV$$

(3.41)

Note that these nonlinear effects occur only in the piezo layers. Inserting$\iota^{NL-FORCE}$ and $\iota^{NL-\sigma}$ in the variational formulation Eq. 3.20, for the smart electro-elastic cantilever beam with a tip load F, gives rise to:

$$\iiint (\sigma_x^L \delta\epsilon_x + \sigma_{xz}^L \delta\gamma_{yz})dV - \iiint (D_x\delta E_x + D_z\delta E_z)dV + \iota^{NL-FORCE} - \iota^{NL-\sigma}$$
$$= F\delta w]_{x=L}$$

(3.42)

Where σ_x^L and σ_{xz}^L are given by Eqs, 3.15 and 3.16 respectively.

3.4.6 FEM Articulation of a Smart Beam

The variational formulation for the Simplified electro-elastic and electro-thermo-elastic smart beam have been obtained in the analysis presented thus far, the generic approximate layer-by-layer representation of the displacement field and the potential have also been given in Eqs.3.8 - 3.11.From the representation given by Eqs. 3.8 - 3.11, it is clear that the displacement field and potential distribution can be obtained if the functions $u_0^l(x),\ u_1^l(x),\ u_2^l(x),\ u_3^l(x),\ \phi_0^l(x),\ \phi_0^l(x),\ \phi_0^l(x)\ and\ w(x)$ are obtained precisely.

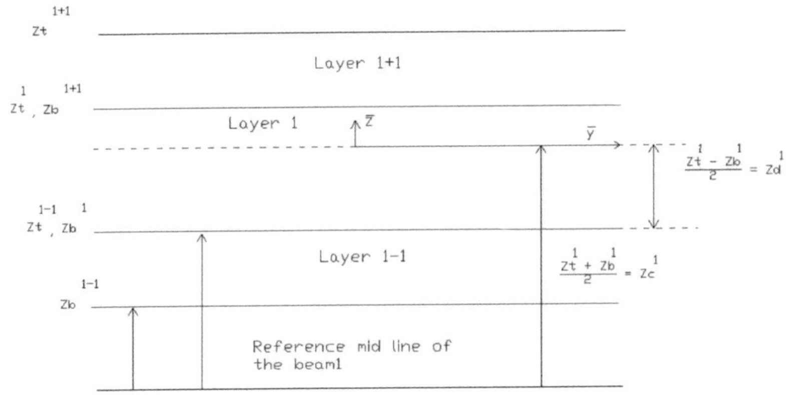

Figure 3.3: Representation of Layers of a Beam

In order to provide a clear description, consider the smart cantilever beam, which is a three-layered beam with outer piezo layers and an inner metallic host. To facilitate the imposition of the inter layer continuity conditions and to incorporate the representation of the displacement and potential fields in a finite element environs, Eqs. 3.8 - 3.10, and Eqs. 3.12 – 3.14, can be revised in terms of the local variable in the thickness direction for each layer, as:

$$u^l(x,z) = \sum_{j=1}^{4} u_j^l\,(x)F_j^l(z)$$

Where $F_j^l(z)$ [1] are written in the transformed variable $\eta = \frac{z-z_c^l}{z_d^l}$ as:

$$F_1^l(z) \ = -\frac{9}{16}(\eta + \frac{1}{3})(\eta - \frac{1}{3})(\eta - 1)$$

$$F_2^l(z) = \frac{27}{16}(\eta + 1)(\eta - \frac{1}{3})(\eta - 1)$$

$$F_3^l(z) = -\frac{27}{16}(\eta + 1)(\eta + \frac{1}{3})(\eta - 1)$$

$$F_4^l(z) = \frac{9}{16}(\eta + 1)(\eta + \frac{1}{3})$$

Where $z_c^l = (z_b^l + z_t^l)/2$ and $z_d^l = (z_t^l - z_b^l)/2$

Similarly the potential $\phi^l(x,z)$is expanded as:

$$\phi^l(x,z) = \sum_{m=1}^{3} \phi_m^l(x) G_m^l(z)$$

Where

$$G_1^l(z) = \frac{1}{2}\eta(\eta - 1)$$

$$G_2^l(z) = -(\eta + 1)(\eta - 1)$$

$$G_3^l(z) = \frac{1}{2}\eta(\eta + 1)$$

Noting that $\eta = \overline{z}/z_d^l$ is scaled local variable $\overline{z} = z - z_c$ such that η = -1 at $z = z_b^l$ and η = 1 at $z = z_t^l$

The lateral deformation $w^l(x,z)$ is taken same for all layers.

Hence $w^l(x,z) = w_0(x)$

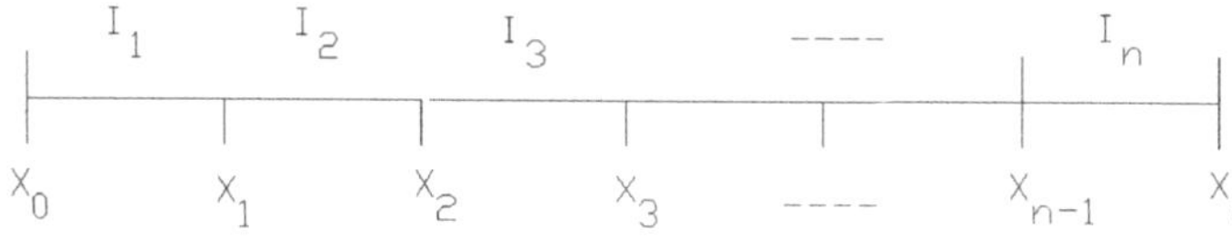

Figure 3.4: Nodes & Elements of a Beam

The functions $u_j^l(x), \phi_m^l(x) and\, w_0(x)$ cannot be obtained exactly in general. Hence, these functions will be approximated using FEM. For the application of the finite element method, the beam will be divided axially into N elements with each element I_j (shown in Fig. 3.4) consisting of three layers. From the variational formulation for linear electro-elastic beams for different cases as given by Eq. 3.21, Eq. 3.28 and Eq.3.36, it can be observed that only the first derivatives of $u_j^l(x),\ \phi_m^l(x)\ and\ w_0(x)$ are present in the integrand. Thus, to ensure that all the terms in the variational formulation are finite, the minimum smoothness requirement is that all the values of $\frac{\partial u_j^l}{\partial x},\ \frac{\partial \phi_m^l}{\partial x}\ and\ \frac{\partial w_0}{\partial x}$ are bounded. This requires that the approximating functions be continuous throughout the length L. Thus, for the finite element approximation C° functions can be used to approximate $u_j^l(x)$, $\phi_m^l(x)\ and\ w_0(x)$. In the present study elementwise cubic Lagrangian basis functions are used to approximate $u_j^l(x),\ \phi_m^l(x)\ and\ w_0(x)$. The approximate functions are represented as:

$$u_j^l(x) \approx \overline{u}_j^l(x) = \sum_{k=1}^{(3N+1)} u_{jk}^l\, \psi_k(x), \qquad j\ varies\ from 1 to 4$$

$$\phi_m^l(x) \approx \overline{\phi}_j^l(x) = \sum_{k=1}^{(3N+1)} u_{jk}^l\, \psi_k(x), \qquad j = 1 to\ 4$$

$$w_0(x) \approx \overline{w}(x) = \sum_{k=1}^{(3N+1)} w_k\, \psi_k(x)$$

Where (3N+1) are the total number of unknown coefficients due to N cubic elements; $\overline{u}_j^l(x),\ \overline{\phi}_j^l(x)$ and $\overline{w}(x)$ are the finite element approximations of $u_j^l(x)$, $\phi_m^l(x)\ and\ w_0(x)$ respectively.

From the continuity condition for the displacement and electric potential between any two layers l and l+1, one obtains

$$\overline{u}_4^{l}(x) = \overline{u}_1^{l+1}(x) \; and \; \overline{\phi}_3^{l}(x) = \overline{\phi}_1^{l+1}(x)$$

It may be noted that ϕ is defined only in the piezo layers, hence no interlayer continuity condition needs to be imposed for interfaces between ferroelectric and metallic layers.

Following standard finite element convention, in an element I_j , the functions $\psi_k(x)$, are represented in terms of the four cubic Lagrangian shape functions $F_i(\xi)$ with ξ as the coordinate over the standard master element $\hat{I}$. For a given element I_j, the transformed coordinate ξ is related to the axial coordinate x by

$$x = x_1^{I_j}\frac{(1-\xi)}{2} + x_2^{I_j}\frac{(1+\xi)}{2}$$

Where $x_1^{I_j}$ and $x_2^{I_j}$ are the end vertices of element I_j and ξ varies from -1 to +1.

In terms of the element shape functions $N_i(\xi)$, the finite element solution in element I_k is given as:

$$\overline{u}_i^{l,I_j}(x) = \sum_{k=1}^{4} u_{ik}^{l,I_j} N_k\{\xi(x)\}$$

$$\overline{\phi}_i^{l,I_j}(x) = \sum_{k=1}^{4} \phi_{ik}^{l,I_j} N_k\{\xi(x)\}$$

And transverse displacement

$$w^{I_j}(x) = \sum_{k=1}^{4} u_k^{I_j} N_k\{\xi(x)\}$$

Where u_{ik}^{l,I_j}, ϕ_{ik}^{l,I_j} and $w^{I_j}(x)$ are the nodal degrees of freedom in I_j -th element in the l - th layer.

The Elemental Stiffness Matrix

For an element I_j there are two types of layers, namely, piezo and metallic. For computing the elemental stiffness and load vectors, the offering from ferroelectric layer and metallic centre host layer will be procured individually. The nodal DOF for an element in the piezo layer material and an element residing in layer of central host metal are shown in Figures.

Piezo Layer

For the piezo layer the functions that need to be considered are

four $u_j^l(x), three\ \phi_m^l(x)\ and\ one\ \ w_0(x).$

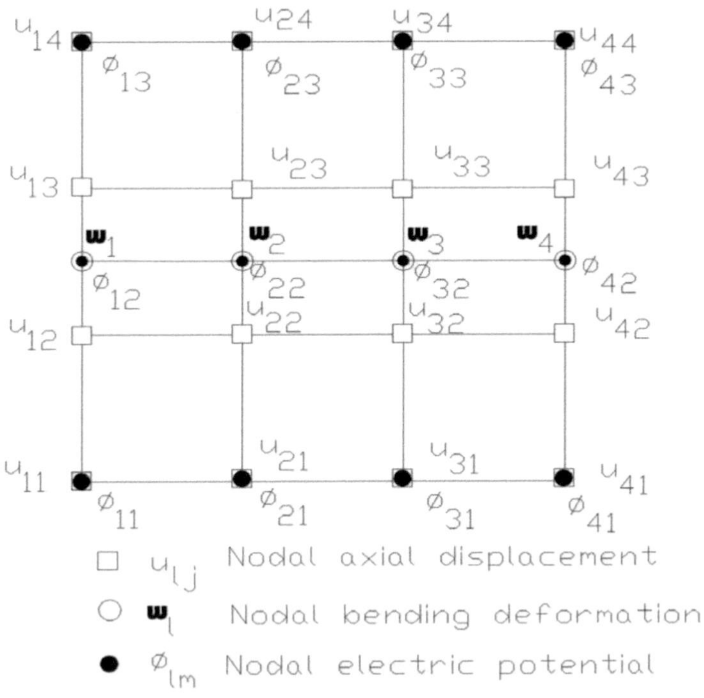

Figure 3.5: FEM Representation of an Element from Piezo Layer

The stiffness for ferroelectric layer is 32×32 matrix, the pure elastic stiffness matrix has an order of 20×20 and an electro elastic stiffness matrix has an order of 20×12. The pure electric stiffness matrix has as order of 12×12 based on FEM approximations listed.

$$\begin{Bmatrix} (K_{11}^{mechanical})_{20\times20} & (K_{12}^{electro-mechanical})_{20\times12} \\ (K_{21}^{electro-elastic})_{12\times20} & (K_{11}^{electric})_{12\times12} \end{Bmatrix} \tag{3.43}$$

The non-piezo layer is non-polarised medium and has an order of 20×20 .The DOF of an element in the stiffness matrix has been given as:

$$\begin{Bmatrix} \left\{ \begin{Bmatrix} axial\ displacement(u_{ij}) \\ transverse\ displacement(w_j) \end{Bmatrix}_{i=1\ to\ 4} \right\}_{j=1\ to\ 4} \\ \left\{ \{electric\ potential(\phi_{ij})\}_{i=1\ to\ 4} \right\}_{j=1 to\ 4} \end{Bmatrix} \tag{3.44}$$

Stiffness matrix is composed of four separate parts corresponding to mechanical elastic phenomenon, electro-mechanical elastic interaction and electro phenomenon. The component of the stiffness are procured by including the elementwise illustration of the FEM solutions in the variational formulation given by Eqs. (3.21) or (3.29) or (3.36). (Representing different cases of study), finally integrating the equations over the layer.

Central Metal Host Layer

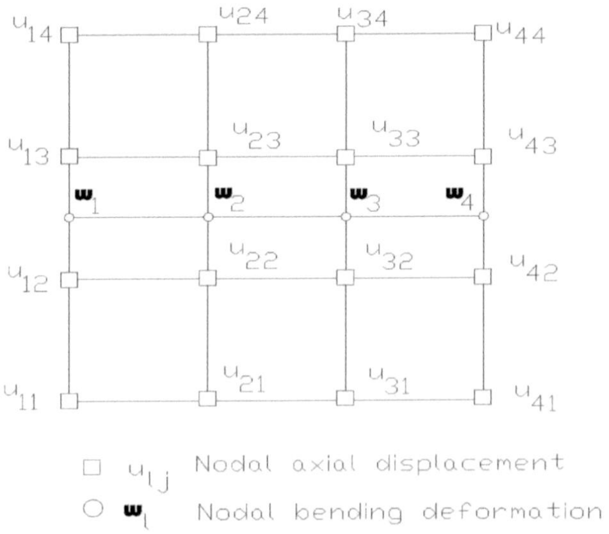

Figure 3.6: FEM Representation of an Element from Metallic Host Layer

The absence of electro phenomenon in central metal host layer ensures only deformations that are possible are elastic in nature, hence for an element I_l in this type of layer only possible elastic deformation are axial deformation $u_j^l(x)$ and transverse deformation $w_0(x)$ generating a stiffness matrix of order 20×20 The component of the stiffness and load vector are procured by including the elementwise illustration of the FEM solutions in the variational formulation given by Eqs. 3.21 or 3.29 or 3.36. (Representing different cases of study), finally integrating the equations over the layer. The elemental stiffness matrix and load vectors are obtained by assembling the layer wise contributions. Note that once the layer wise elemental contributions to the stiffness matrix and the load vector are obtained separately for the piezo and the metallic layers, the final element stiffness matrix and load vector can be obtained easily for any stacking sequence of the layers, i.e. for the surface mounted beam or the sandwich beam structure.

3.5 Computation of the Global stiffness matrix [K] and load vector {F} for the smart beam

The global stiffness matrix [K], global degrees of freedom matrix {a} and load vector {F} for the smart beam are related as:

$$[K]_{m\times m}\,\{a\}_{m\times 1} = \{F\}_{m\times 1} \tag{3.45}$$

Where m represent number of global degrees of freedom.

$[K]_{m\times m}$ and $\{F\}_{m\times 1}$ are procured for cantilever beam by assembling K matrices and F vectors.

Where K is the elemental stiffness matrix and F is the elemental load vector. For the nonlinear formulation, the finite element problem has to be solved iteratively. Using the variational formulation given by Eq. 3.42, the global stiffness matrix and load vector for the nonlinear problem can be written as:

$$[K]\{a\} + [K_1(u,w,\phi)]\{a\} + [K_2(u,w,\phi)]\{a\} = \{F\}$$

Where $[K_1(u,w,\phi)]$ *and* $[K_2(u,w,\phi)]$ represent the non-linear extended stiffness matrices due to distributed nonlinear force and nonlinear stress terms respectively.

Summary

Variation form of the governing equations has been presented by invoking principal of virtual work and appropriate equations boundary conditions identified. Constitutive equations have been reduced for certain simplified cases. A layer by layer FEM modeling of the smart beam has been developed to properly account for the material discontinuity across the thickness of the smart cantilever beam.

References:

[01] Paolo Gaudenzi "Smart Structures Physical Behaviour, Mathematical Modelling and Applications" John Wiley Publications

[02] Cady, W.G., Piezoelectricity, An Introduction to the Theory and Applications of Electro-mechanical Phenomena in Crystals, 1st Ed. Vols. 1 and 2, Dover Publications, INC., New York, 1946.

[03] Ikeda, T., Fundamentals of Piezoelectricity, 1st Ed. Oxford University Press, 1990.

[04] Sateesh.V,L., Upadhyay, C,S., and Venkatesan,C., "Thermodynamic modeling of Hysteresis effects in piezoceramics for application to smart structures" AIAA Journal 46(1), 2008.

[05] Landau, L. D., Lifshitz, E. M., and Pitaevski ı ˘ , L. P., *Electrodynamics of Continuous Media*, 2nd ed., Vol. 8, Pergamon, New York, pp. 1-85, 1984.

[06] Hellen, E., *Electromagnetic Theory*, 1st ed. John Wiley & Sons Inc., New York, , pp. 1-189, 1962.

[07] Griffiths, D. J., *Introduction to Electrodynamics*, 3rd ed., Prentice-Hall, India, pp. 160-358, 1999.

[08] Tiersten, H. F., *Linear Piezoelectric Plate Vibrations, Elements of the Linear Theory of Piezoelectricity and the Vibrations of Piezoelectric Plates, 1st ed.*, Plenum Press, New York, 1969.

[09] Gurtin, M. E., *Mathematics in Science and Engineering, An Introduction to Continuum Mechanics,* 1st ed. Vol. 158, Academic Press, New York, 1981.

[11] Malvern, L. E., *Introduction to the Mechanics of a Continuous Medium*, 1st ed., Prentice-Hall, Inc. New Jersey, 1969.

[12] Landau, L. D., and Lifshitz, E. M., *Theory of Elasticity*, 2nd ed., Pergamon, New York, 1970, pp. 1-43.

[13] Chandrasekharaiah, D. S., and Debnath, L. *Continuum Mechanics*, Prism Books PVT Ltd., Bangalore, India, Vol. 7, 1st ed,1994.

[14] Tiersten, H. F., "Coupled Magnetomechanical Equations for Magnetically Saturated Insulators," Journal of Mathematical Physics, Vol. 5, No. 9, pp. 1298-1318, Sept. 1964.

[15] Tiersten, H.F., "On the non-linear Equations of Electro-thermo-elasticity," International Journal of Engineering Sciences, Vol. 9, No. 7, pp. 587-604, 1971.

[16] Ahmad, Sheikh N., Upadhyay, C,S., and Venkatesan,C., "Electro-thermo-Elastic formulation for the analysis of smart structures" Journal of Smart Materials and Structures, No. 15, pp. 401-416, 2006.

[17] Reddy, J. N., *An Introduction to the Finite Element Method*, 3rd ed. McGraw-Hill, New York, 2005

[18] Segerlind, Larry, J.,Applied Finite element analysis John Wiley

Druck:
Canon Deutschland Business Services GmbH
im Auftrag der KNV-Gruppe
Ferdinand-Jühlke-Str. 7
99095 Erfurt